Navigating Mobile Robots

Navigating Mobile Robots

Systems and Techniques

Johann Borenstein
The University of Michigan
Ann Arbor, Michigan

H.R. Everett
Naval Command, Control and Ocean Surveillance Center
San Diego, California

Liqiang Feng
The University of Michigan
Ann Arbor, Michigan

A K Peters
Wellesley, Massachusetts

Editorial, Sales, and Customer Service Office

A K Peters, Ltd.
289 Linden Street
Wellesley, MA 02181

Library of Congress Cataloging-in-Publication Data

Borenstein, J. (Johann)
    Navigating mobile robots : systems and techniques / Johann
Borenstein, H.R. Everett, Liqiang Feng.
      p. cm.
    Includes bibliographical references and index.
    ISBN 1-56881-058-X
    1. Mobile robots. 2. Robots — Control systems. I. Everett, H.
R., 1949-   II. Feng, L. (Liqiang) III. Title.
TJ211.415.B67 1996   96-4936
629.8'92—dc20   CIP

Cover photograph reprinted with permission from
David Kother, Ann Arbor, Michigan.

Printed in the United States of America
00 99 98 97 96   10 9 8 7 6 5 4 3 2 1

Acknowledgments

The authors wish to thank the Department of Energy (DOE), and especially
Dr.Linton W. Yarbrough, DOE Program Manager, Dr. William R. Hamel, D&D
Technical Coordinator, and Dr. Clyde Ward, Landfill Operations Technical
Coordinator for their technical and financial support of the
research, which forms the basis of this work.
Other significant portions of the text were adapted from *Sensors for Mobile
Robots: Theory and Application*, by H. R. Everett, A K Peters, Ltd.,
Wellesley, MA, Publishers.

The authors further wish to thank Professors David K. Wehe and Yoram Koren
at the University of Michigan for their support, and Mr. Harry Alter (DOE)
who has befriended many of the graduate students and sired several of our robots.

Thanks are also due to Todd Ashley Everett for making most of the line-art drawings.

Table of Contents

INTRODUCTION

Leonard and Durrant-Whyte [1991] summarized the general problem of mobile robot navigation by three questions: "Where am I?," "Where am I going?," and "How should I get there?." This book surveys the state-of-the-art in sensors, systems, methods, and technologies that aim at answering the first question, that is: robot positioning in its environment.

Perhaps the most important result from surveying the vast body of literature on mobile robot positioning is that to date there is no truly elegant solution for the problem. The many partial solutions can roughly be categorized into two groups: *relative* and *absolute* position measurements. Because of the lack of a single, generally good method, developers of *automated guided vehicles* (AGVs) and mobile robots usually combine two methods, one from each category. The two categories can be further divided into the following subgroups.

Relative Position Measurements

a. **Odometry** This method uses encoders to measure wheel rotation and/or steering orientation. Odometry has the advantage that it is totally self-contained, and it is always capable of providing the vehicle with an estimate of its position. The disadvantage of odometry is that the position error grows without bound unless an independent reference is used periodically to reduce the error [Cox, 1991].

b. **Inertial Navigation** This method uses gyroscopes and sometimes accelerometers to measure rate of rotation and acceleration. Measurements are integrated once (or twice) to yield position. Inertial navigation systems also have the advantage that they are self-contained. On the downside, inertial sensor data drifts with time because of the need to integrate rate data to yield position; any small constant error increases without bound after integration. Inertial sensors are thus unsuitable for accurate positioning over an extended period of time. Another problem with inertial navigation is the high equipment cost. For example, highly accurate gyros, used in airplanes, are inhibitively expensive. Very recently fiber-optic gyros (also called laser gyros), which are said to be very accurate, have fallen dramatically in price and have become a very attractive solution for mobile robot navigation.

Absolute Position Measurements

c. **Active Beacons** This method computes the absolute position of the robot from measuring the direction of incidence of three or more actively transmitted beacons. The transmitters, usually using light or radio frequencies, must be located at known sites in the environment.

d. **Artificial Landmark Recognition** In this method distinctive artificial landmarks are placed at known locations in the environment. The advantage of artificial landmarks is that they can be designed for optimal detectability even under adverse environmental conditions. As with active beacons, three or more landmarks must be "in view" to allow position estimation. Landmark positioning has the advantage that the position errors are bounded, but detection of external landmarks and real-time position fixing may not always be possible. Unlike the usually point-shaped beacons, artificial landmarks may be defined as a set of features, e.g., a shape or an area. Additional information, for example distance, can be derived from measuring the geometric properties of the landmark, but this approach is computationally intensive and not very accurate.

e. **Natural Landmark Recognition** Here the landmarks are distinctive features in the environment. There is no need for preparation of the environment, but the environment must be known in advance. The reliability of this method is not as high as with artificial landmarks.

f. **Model Matching** In this method information acquired from the robot's onboard sensors is compared to a map or world model of the environment. If features from the sensor-based map and the world model map match, then the vehicle's absolute location can be estimated. Map-based positioning often includes improving global maps based on the new sensory observations in a dynamic environment and integrating local maps into the global map to cover previously unexplored areas. The maps used in navigation include two major types: geometric maps and topological maps. Geometric maps represent the world in a global coordinate system, while topological maps represent the world as a network of nodes and arcs.

This book presents and discusses the state-of-the-art in each of the above six categories. The material is organized in two parts: Part I deals with the sensors used in mobile robot positioning, and Part II discusses the methods and techniques that make use of these sensors.

Mobile robot navigation is a very diverse area, and a useful comparison of different approaches is difficult because of the lack of commonly accepted test standards and procedures. The research platforms used differ greatly and so do the key assumptions used in different approaches. Further difficulty arises from the fact that different systems are at different stages in their development. For example, one system may be commercially available, while another system, perhaps with better performance, has been tested only under a limited set of laboratory conditions. For these reasons we generally refrain from comparing or even judging the performance of different systems or techniques. Furthermore, we have not tested most of the systems and techniques, so the results and specifications given in this book are merely quoted from the respective research papers or product spec-sheets.

Because of the above challenges we have defined the purpose of this book to be a survey of the expanding field of mobile robot positioning. It took well over 1.5 man-years to gather and compile the material for this book; we hope this work will help the reader to gain greater understanding in much less time.

Part I
Sensors for
Mobile Robot Positioning

CARMEL, the University of Michigan's first mobile robot, has been in service since 1987. Since then, CARMEL has served as a reliable testbed for countless sensor systems. In the extra "shelf" underneath the robot is an 8086 XT compatible single-board computer that runs U of M's ultrasonic sensor firing algorithm. Since this code was written in 1987, the computer has been booting up and running from *floppy disk*. The program was written in FORTH and was never altered; should anything ever go wrong with the floppy, it will take a computer *historian* to recover the code...

Chapter 1
Sensors for Dead Reckoning

Dead reckoning (derived from "deduced reckoning" of sailing days) is a simple mathematical procedure for determining the present location of a vessel by advancing some previous position through known course and velocity information over a given length of time [Dunlap and Shufeldt, 1972]. The vast majority of land-based mobile robotic systems in use today rely on dead reckoning to form the very backbone of their navigation strategy, and like their nautical counterparts, periodically null out accumulated errors with recurring "fixes" from assorted navigation aids.

The most simplistic implementation of dead reckoning is sometimes termed *odometry*; the term implies vehicle displacement along the path of travel is directly derived from some onboard "odometer." A common means of odometry instrumentation involves optical encoders directly coupled to the motor armatures or wheel axles.

Since most mobile robots rely on some variation of wheeled locomotion, a basic understanding of sensors that accurately quantify angular position and velocity is an important prerequisite to further discussions of odometry. There are a number of different types of rotational displacement and velocity sensors in use today:
- Brush encoders.
- Potentiometers.
- Synchros.
- Resolvers.
- Optical encoders.
- Magnetic encoders.
- Inductive encoders.
- Capacitive encoders.

A multitude of issues must be considered in choosing the appropriate device for a particular application. Avolio [1993] points out that over 17 million variations on rotary encoders are offered by one company alone. For mobile robot applications incremental and absolute optical encoders are the most popular type. We will discuss those in the following sections.

1.1 Optical Encoders

The first optical encoders were developed in the mid-1940s by the Baldwin Piano Company for use as "tone wheels" that allowed electric organs to mimic other musical instruments [Agent, 1991]. Today's corresponding devices basically embody a miniaturized version of the *break-beam proximity sensor*. A focused beam of light aimed at a matched photodetector is periodically interrupted by a coded opaque/transparent pattern on a rotating intermediate disk attached to the shaft of interest. The rotating disk may take the form of chrome on glass, etched metal, or *photoplast* such as Mylar [Henkel, 1987]. Relative to the more complex alternating-current resolvers, the straightforward encoding scheme and inherently digital output of the optical encoder results in a low-cost reliable package with good noise immunity.

There are two basic types of optical encoders: *incremental* and *absolute*. The incremental version measures rotational velocity and can infer relative position, while absolute models directly measure angular position and infer velocity. If non volatile position information is not a consideration, *incremental encoders* generally are easier to interface and provide equivalent resolution at a much lower cost than *absolute* optical encoders.

1.1.1 Incremental Optical Encoders

The simplest type of incremental encoder is a single-channel *tachometer encoder,* basically an instrumented mechanical light chopper that produces a certain number of sine- or square-wave pulses for each shaft revolution. Adding pulses increases the resolution (and subsequently the cost) of the unit. These relatively inexpensive devices are well suited as velocity feedback sensors in medium- to high-speed control systems, but run into noise and stability problems at extremely slow velocities due to quantization errors [Nickson, 1985]. The tradeoff here is resolution versus update rate: improved transient response requires a faster update rate, which for a given line count reduces the number of possible encoder pulses per sampling interval. A very simple, do-it-yourself encoder is described in [Jones and Flynn, 1993]. More sophisticated single-channel encoders are typically limited to 2540 lines for a 5-centimeter (2 in) diameter incremental encoder disk [Henkel, 1987].

In addition to low-speed instabilities, single-channel tachometer encoders are also incapable of detecting the direction of rotation and thus cannot be used as position sensors. *Phase-quadrature incremental encoders* overcome these problems by adding a second channel, displaced from the first, so the resulting pulse trains are 90 degrees out of phase as shown in Figure 1.1. This technique allows the decoding electronics to determine which channel is leading the other and hence ascertain the direction of rotation, with the added benefit of increased resolution. Holle [1990] provides an in-depth discussion of output options (single-ended TTL or differential drivers) and various design issues (i.e., resolution, bandwidth, phasing, filtering) for consideration when interfacing phase-quadrature incremental encoders to digital control systems.

The incremental nature of the phase-quadrature output signals dictates that any resolution of angular position can only be relative to some specific reference, as opposed to absolute. Establishing such a reference can be accomplished in a number of ways. For applications involving

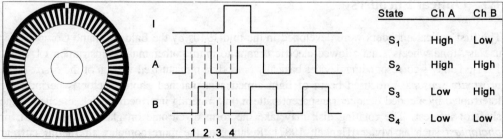

Figure 1.1: The observed phase relationship between Channel A and B pulse trains can be used to determine the direction of rotation with a phase-quadrature encoder, while unique output states S_1 - S_4 allow for up to a four-fold increase in resolution. The single slot in the outer track generates one index pulse per disk rotation [Everett, 1995].

continuous 360-degree rotation, most encoders incorporate as a third channel a special *index output* that goes high once for each complete revolution of the shaft (see Figure 1.1 above). Intermediate shaft positions are then specified by the number of encoder up counts or down counts from this known index position. One disadvantage of this approach is that all relative position information is lost in the event of a power interruption.

In the case of limited rotation, such as the back-and-forth motion of a pan or tilt axis, electrical limit switches and/or mechanical stops can be used to establish a home reference position. To improve repeatability this homing action is sometimes broken into two steps. The axis is rotated at reduced speed in the appropriate direction until the stop mechanism is encountered, whereupon rotation is reversed for a short predefined interval. The shaft is then rotated slowly back into the stop at a specified low velocity from this designated start point, thus eliminating any variations in inertial loading that could influence the final homing position. This two-step approach can usually be observed in the power-on initialization of stepper-motor positioners for dot-matrix printer heads.

Alternatively, the absolute indexing function can be based on some external referencing action that is decoupled from the immediate servo-control loop. A good illustration of this situation involves an incremental encoder used to keep track of platform steering angle. For example, when the *K2A Navmaster* [CYBERMOTION] robot is first powered up, the absolute steering angle is unknown, and must be initialized through a "referencing" action with the docking beacon, a nearby wall, or some other identifiable set of landmarks of known orientation. The up/down count output from the decoder electronics is then used to modify the vehicle heading register in a relative fashion.

A growing number of very inexpensive off-the-shelf components have contributed to making the phase-quadrature incremental encoder the rotational sensor of choice within the robotics research and development community. Several manufacturers now offer small DC gear-motors with incremental encoders already attached to the armature shafts. Within the U.S. automated guided vehicle (AGV) industry, however, resolvers are still generally preferred over optical encoders for their perceived superiority under harsh operating conditions, but the European AGV community seems to clearly favor the encoder [Manolis, 1993].

Interfacing an incremental encoder to a computer is not a trivial task. A simple state-based interface as implied in Figure 1.1 is inaccurate if the encoder changes direction at certain positions, and false pulses can result from the interpretation of the sequence of state changes [Pessen, 1989]. Pessen describes an accurate circuit that correctly interprets directional state changes. This circuit was originally developed and tested by Borenstein [1987].

A more versatile encoder interface is the HCTL 1100 motion controller chip made by Hewlett Packard [HP]. The HCTL chip performs not only accurate quadrature decoding of the incremental wheel encoder output, but it provides many important additional functions, including among others:
- closed-loop position control,
- closed-loop velocity control in P or PI fashion,
- 24-bit position monitoring.

At the University of Michigan's Mobile Robotics Lab, the HCTL 1100 has been tested and used in many different mobile robot control interfaces. The chip has proven to work reliably and accurately, and it is used on commercially available mobile robots, such as the TRC *LabMate* and *HelpMate*. The HCTL 1100 costs only $40 and it comes highly recommended.

1.1.2 Absolute Optical Encoders

Absolute encoders are typically used for slower rotational applications that require positional information when potential loss of reference from power interruption cannot be tolerated. Discrete detector elements in a photovoltaic array are individually aligned in break-beam fashion with concentric encoder tracks as shown in Figure 1.2, creating in effect a non-contact implementation of a commutating brush encoder. The assignment of a dedicated track for each bit of resolution results in larger size disks (relative to incremental designs), with a corresponding decrease in shock and vibration tolerance. A general rule of thumb is that each additional encoder track doubles the resolution but quadruples the cost [Agent, 1991].

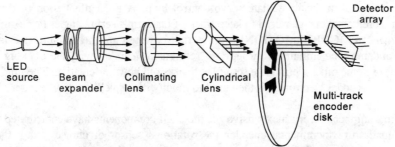

Figure 1.2: A line source of light passing through a coded pattern of opaque and transparent segments on the rotating encoder disk results in a parallel output that uniquely specifies the absolute angular position of the shaft. (Adapted from [Agent, 1991].)

Instead of the serial bit streams of incremental designs, absolute optical encoders provide a parallel word output with a unique code pattern for each quantized shaft position. The most common coding schemes are Gray code, natural binary, and binary-coded decimal [Avolio, 1993]. The Gray code (for inventor Frank Gray of Bell Labs) is characterized by the fact that only one bit changes at a time, a decided advantage in eliminating asynchronous ambiguities caused by electronic and mechanical component tolerances (see Figure 1.3a). Binary code, on the other hand, routinely involves multiple bit changes when incrementing or decrementing the count by one. For example, when going from position 255 to position 0 in Figure 1.3b, eight bits toggle from 1s to 0s. Since there is no guarantee all threshold detectors monitoring the detector elements tracking each bit will toggle at the same precise instant, considerable ambiguity can exist during state transition with a coding scheme of this form. Some type of handshake line signaling valid data available would be required if more than one bit were allowed to change between consecutive encoder positions.

Absolute encoders are best suited for slow and/or infrequent rotations such as steering angle encoding, as opposed to measuring high-speed continuous (i.e., drive wheel) rotations as would be required for calculating displacement along the path of travel. Although not quite as robust as resolvers for high-temperature, high-shock applications, absolute encoders can operate at temperatures over 125°C, and medium-resolution (1000 counts per revolution) metal or Mylar disk designs can compete favorably with resolvers in terms of shock resistance [Manolis, 1993]. A potential disadvantage of absolute encoders is their parallel data output, which requires a more complex interface due to the large number of electrical leads. A 13-bit absolute encoder using complimentary output signals for noise immunity would require a 28-conductor cable (13 signal pairs plus power and ground), versus only six for a resolver or incremental encoder [Avolio, 1993].

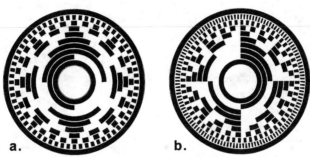

Figure 1.3: Rotating an 8-bit absolute Gray code disk.
a. Counterclockwise rotation by one position increment will cause only one bit to change.
b. The same rotation of a binary-coded disk will cause all bits to change in the particular case (255 to 0) illustrated by the reference line at 12 o'clock.
[Everett, 1995].

1.2 Doppler Sensors

The rotational displacement sensors discussed above derive navigation parameters directly from wheel rotation, and are thus subject to problems arising from slippage, tread wear, and/or improper tire inflation. In certain applications, Doppler and inertial navigation techniques are sometimes employed to reduce the effects of such error sources.

Doppler navigation systems are routinely employed in maritime and aeronautical applications to yield velocity measurements with respect to the earth itself, thus eliminating dead-reckoning errors introduced by unknown ocean or air currents. The principle of operation is based on the Doppler shift in frequency observed when radiated energy reflects off a surface that is moving with respect to the emitter. Maritime systems employ acoustical energy reflected from the ocean floor, while airborne systems sense microwave RF energy bounced off the surface of the earth. Both configurations typically involve an array of four transducers spaced 90 degrees apart in azimuth and inclined downward at a common angle with respect to the horizontal plane [Dunlap and Shufeldt, 1972].

Due to cost constraints and the reduced likelihood of transverse drift, most robotic implementations employ but a single forward-looking transducer to measure ground speed in the direction of travel. Similar configurations are sometimes used in the agricultural industry, where tire slippage in soft freshly plowed dirt can seriously interfere with the need to release seed or fertilizer at a rate commensurate with vehicle advance. The M113-based Ground Surveillance Vehicle [Harmon, 1986] employed an off-the-shelf unit of this type manufactured by John Deere to compensate for track slippage.

The microwave radar sensor is aimed downward at a prescribed angle (typically 45°) to sense ground movement as shown in Figure 1.4. Actual ground speed V_A is derived from the measured velocity V_D according to the following equation [Schultz, 1993]:

$$V_A = \frac{V_D}{\cos\alpha} = \frac{cF_D}{2F_0\cos\alpha} \qquad (1.1)$$

where
V_A = actual ground velocity along path
V_D = measured Doppler velocity
α = angle of declination
c = speed of light
F_D = observed Doppler shift frequency
F_0 = transmitted frequency.

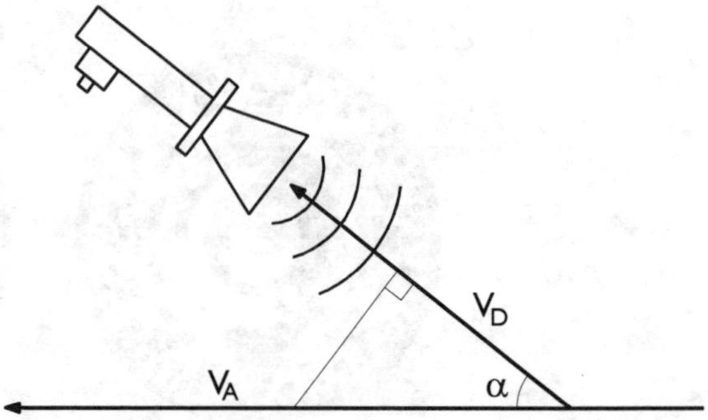

Errors in detecting true ground speed arise due to side-lobe interference, vertical velocity components introduced by vehicle

Figure 1.4: A Doppler ground-speed sensor inclined at an angle α as shown measures the velocity component V_D of true ground speed V_A. (Adapted from [Schultz, 1993].)

reaction to road surface anomalies, and uncertainties in the actual angle of incidence due to the finite width of the beam. Byrne et al. [1992] point out another interesting scenario for potentially erroneous operation, involving a stationary vehicle parked over a stream of water. The Doppler ground-speed sensor in this case would misinterpret the relative motion between the stopped vehicle and the running water as vehicle travel.

1.2.1 Micro-Trak *Trak-Star* Ultrasonic Speed Sensor

One commercially available speed sensor that is based on Doppler speed measurements is the *Trak-Star* Ultrasonic Speed Sensor [MICRO-TRAK]. This device, originally designed for agricultural applications, costs $420. The manufacturer claims that this is the most accurate Doppler speed sensor available. The technical specifications are listed in Table 1.1.

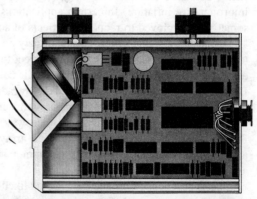

Figure 1.5: The *Trak-Star* Ultrasonic Speed Sensor is based on the Doppler effect. This device is primarily targeted at the agricultural market. (Courtesy of Micro-Trak.)

1.2.2 Other Doppler-Effect Systems

A non-radar Doppler-effect device is the *Monitor 1000*, a distance and speed monitor for runners. This device was temporarily marketed by the sporting goods manufacturer [NIKE]. The *Monitor 1000* was worn by the runner like a front-mounted fanny pack. The small and lightweight device used ultrasound as the carrier, and was said to have an accuracy of two to five percent, depending on the ground characteristics. The manufacturer of the *Monitor 1000* is Applied Design Laboratories [ADL]. A microwave radar Doppler effect distance sensor has also been developed by ADL. This radar sensor is a prototype and is not commercially available. However, it differs from the *Monitor 1000* only in its use of

Table 1.1: Specifications for the *Trak-Star* Ultrasonic Speed Sensor.

Parameter	Value	Units
Speed range	17.7	m/s
	0-40	mph
Speed resolution	1.8	cm/s
	0.7	in/s
Accuracy	±1.5%+0.04	mph
Transmit frequency	62.5	kHz
Temperature range	-29 to +50	°C
	-20 to +120	°F
Weight	1.3	kg
	3	lb
Power requirements	12	VDC
	0.03	A

a radar sensor head as opposed to the ultrasonic sensor head used by the *Monitor 1000*. The prototype radar sensor measures $15 \times 10 \times 5$ centimeters ($6 \times 4 \times 2$ in), weighs 250 grams (8.8 oz), and consumes 0.9 W

1.3   Typical Mobility Configurations

The accuracy of odometry measurements for dead reckoning is to a great extent a direct function of the kinematic design of a vehicle. Because of this close relation between kinematic design and positioning accuracy, one must consider the kinematic design closely before attempting to improve dead-reckoning accuracy. For this reason, we will briefly discuss some of the more popular vehicle designs in the following sections. In Part II of this report, we will discuss some recently developed methods for reducing odometry errors (or the feasibility of doing so) for some of these vehicle designs.

1.3.1 Differential Drive

Figure 1.6 shows a typical *differential drive* mobile robot, the *LabMate* platform, manufactured by [TRC]. In this design incremental encoders are mounted onto the two drive motors to count the wheel revolutions. The robot can perform dead reckoning by using simple geometric equations to compute the momentary position of the vehicle relative to a known starting position.

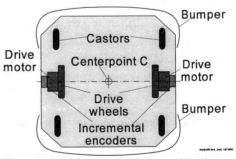

Figure 1.6: A typical differential-drive mobile robot (bottom view).

For completeness, we rewrite the well-known equations for odometry below (also, see [Klarer, 1988; Crowley and Reignier, 1992]). Suppose that at sampling interval I the left and right wheel encoders show a pulse increment of N_L and N_R, respectively. Suppose further that

$$c_m = \pi D_n/nC_e \tag{1.2}$$

where
c_m = conversion factor that translates encoder pulses into linear wheel displacement
D_n = nominal wheel diameter (in mm)
C_e = encoder resolution (in pulses per revolution)
n = gear ratio of the reduction gear between the motor (where the encoder is attached) and the drive wheel.

We can compute the incremental travel distance for the left and right wheel, $\Delta U_{L,i}$ and $\Delta U_{R,i}$, according to

$$\Delta U_{L/R, i} = c_m\, N_{L/R, i} \tag{1.3}$$

and the incremental linear displacement of the robot's centerpoint C, denoted ΔU_i, according to

$$\Delta U_i = (\Delta U_R + \Delta U_L)/2. \tag{1.4}$$

Next, we compute the robot's incremental change of orientation

$$\Delta\theta_i = (\Delta U_R - \Delta U_L)/b \tag{1.5}$$

where b is the wheelbase of the vehicle, ideally measured as the distance between the two contact points between the wheels and the floor.

The robot's new relative orientation θ_i can be computed from

$$\theta_i = \theta_{i-1} + \Delta\theta_i \tag{1.6}$$

and the relative position of the centerpoint is

$$x_i = x_{i-1} + \Delta U_i \cos\theta_i \tag{1.7a}$$
$$y_i = y_{i-1} + \Delta U_i \sin\theta_i \tag{1.7b}$$

where
x_i, y_i = relative position of the robot's centerpoint c at instant i.

1.3.2 Tricycle Drive

Tricycle-drive configurations (see Figure 1.7) employing a single driven front wheel and two passive rear wheels (or vice versa) are fairly common in AGV applications because of their inherent simplicity. For odometry instrumentation in the form of a steering-angle encoder, the dead-reckoning solution is equivalent to that of an Ackerman-steered vehicle, where the steerable wheel replaces the imaginary center wheel discussed in Section 1.3.3. Alternatively, if rear-axle differential odometry is used to determine heading, the solution is identical to the differential-drive configuration discussed in Section 1.3.1.

One problem associated with the tricycle-drive configuration is that the vehicle's center of gravity tends to move away from the front wheel when traversing up an incline, causing a loss of traction. As in the case of Ackerman-steered designs, some surface damage and induced heading errors are possible when actuating the steering while the platform is not moving.

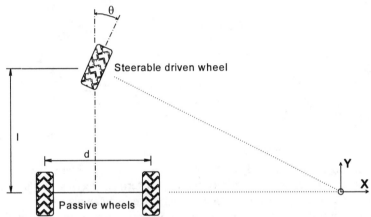

Figure 1.7: Tricycle-drive configurations employing a steerable driven wheel and two passive trailing wheels can derive heading information directly from a steering angle encoder or indirectly from differential odometry [Everett, 1995].

1.3.3 Ackerman Steering

Used almost exclusively in the automotive industry, Ackerman steering is designed to ensure that the inside front wheel is rotated to a slightly sharper angle than the outside wheel when turning, thereby eliminating geometrically induced tire slippage. As seen in Figure 1.8, the extended axes for the two front wheels intersect in a common point that lies on the extended axis of the rear axle. The locus of points traced along the ground by the center of each tire is thus a set of concentric arcs about this centerpoint of rotation P_1, and (ignoring for the moment any centrifugal accelerations) all instantaneous velocity vectors will subsequently be tangential to these arcs. Such a steering geometry is said to satisfy the Ackerman equation [Byrne et al., 1992]:

$$\cot\theta_i - \cot\theta_o = \frac{d}{l} \tag{1.8}$$

where
θ_i = relative steering angle of the inner wheel
θ_o = relative steering angle of the outer wheel
l = longitudinal wheel separation
d = lateral wheel separation.

For the sake of convenience, the vehicle steering angle θ_{SA} can be thought of as the angle (relative to vehicle heading) associated with an imaginary center wheel located at a reference point P_2 as shown in the figure above. θ_{SA} can be expressed in terms of either the inside or outside steering angles (θ_i or θ_o) as follows [Byrne et al., 1992]:

$$\cot\theta_{SA} = \frac{d}{2l} + \cot\theta_i \tag{1.9}$$

or, alternatively,

$$\cot\theta_{SA} = \cot\theta_o - \frac{d}{2l} . \tag{1.10}$$

Ackerman steering provides a fairly accurate odometry solution while supporting the traction and ground clearance needs of all-terrain operation. Ackerman steering is thus the method of choice for outdoor autonomous vehicles. Associated drive implementations typically employ a gasoline or diesel engine coupled to a manual or automatic transmission, with power applied to

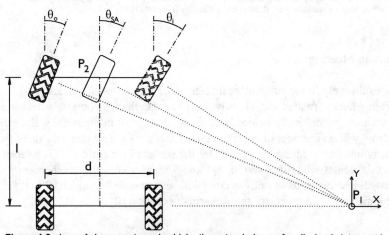

Figure 1.8: In an Ackerman-steered vehicle, the extended axes for all wheels intersect in a common point. (Adapted from [Byrne et al., 1992].)

four wheels through a transfer case, a differential, and a series of universal joints. A representative example is seen in the HMMWV-based prototype of the USMC Tele-Operated Vehicle (TOV) Program [Aviles et al., 1990]. From a military perspective, the use of existing-inventory equipment of this type simplifies some of the logistics problems associated with vehicle maintenance. In addition, reliability of the drive components is high due to the inherited stability of a proven power train. (Significant interface problems can be encountered, however, in retrofitting off-the-shelf vehicles intended for human drivers to accommodate remote or computer control.)

1.3.4 Synchro Drive

An innovative configuration known as *synchro drive* features three or more wheels (Figure 1.9) mechanically coupled in such a way that all rotate in the same direction at the same speed, and similarly pivot in unison about their respective steering axes when executing a turn. This drive and steering "synchronization" results in improved odometry accuracy through reduced slippage, since all wheels generate equal and parallel force vectors at all times.

The required mechanical synchronization can be accomplished in a number of ways, the most common being a chain, belt, or gear drive. Carnegie Mellon University has implemented an electronically synchronized version on one of their *Rover* series robots, with dedicated drive motors for each of the three wheels. Chain- and belt-drive configurations experience some degradation in steering accuracy and alignment due to uneven distribution of slack, which varies as a function of loading and direction of rotation. In addition, whenever chains (or timing belts) are tightened to reduce such slack, the individual wheels must be realigned. These problems are eliminated with a completely enclosed gear-drive approach. An enclosed gear train also significantly reduces noise as well as particulate generation, the latter being very important in clean-room applications.

An example of a three-wheeled belt-drive implementation is seen in the Denning *Sentry* formerly manufactured by Denning Mobile Robots, Woburn, MA [Kadonoff, 1986] and now by Denning Branch Robotics International [DBIR]. Referring to Figure 1.9, drive torque is transferred down through the three steering columns to polyurethane-filled rubber tires. The drive-

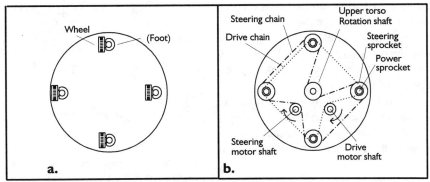

Figure 1.9:   A four-wheel synchro-drive configuration: a. Bottom view.   b. Top view.
(Adapted from Holland [1983].)

motor output shaft is mechanically coupled to each of the steering-column power shafts by a heavy-duty timing belt to ensure synchronous operation. A second timing belt transfers the rotational output of the steering motor to the three steering columns, allowing them to synchronously pivot throughout a full 360-degree range [Everett, 1985]. The Sentry's upper head assembly is mechanically coupled to the steering mechanism in a manner similar to that illustrated in Figure 1.9, and thus always points in the direction of forward travel. The three-point configuration ensures good stability and traction, while the actively driven large-diameter wheels provide more than adequate obstacle climbing capability for indoor scenarios. The disadvantages of this particular implementation include odometry errors introduced by compliance in the drive belts as well as by reactionary frictional forces exerted by the floor surface when turning in place.

To overcome these problems, the Cybermotion *K2A Navmaster* robot employs an enclosed gear-drive configuration with the wheels offset from the steering axis as shown in Figure 1.10 and Figure 1.11. When a foot pivots during a turn, the attached wheel rotates in the appropriate direction to minimize floor and tire wear, power consumption, and slippage. Note that for correct compensation, the miter gear on the wheel axis must be on the opposite side of the power shaft gear from the wheel as illustrated. The governing equation for minimal slippage is [Holland, 1983]

$$\frac{A}{B} = \frac{r'}{r} \tag{1.11}$$

where
A = number of teeth on the power shaft gear
B = number of teeth on the wheel axle gear
r' = wheel offset from steering pivot axis
r = wheel radius.

One drawback of this approach is seen in the decreased lateral stability that results when one wheel is turned in under the vehicle. Cybermotion's improved *K3A* design solves this problem (with an even smaller wheelbase) by incorporating a dual-wheel arrangement on each foot [Fisher et al., 1994]. The two wheels turn in opposite directions in differential fashion as the foot pivots during a turn, but good stability is maintained in the foregoing example by the outward swing of the additional wheel.

The odometry calculations for the synchro drive are almost trivial; vehicle heading is simply derived from the steering-angle encoder, while displacement in the direction of travel is given as follows:

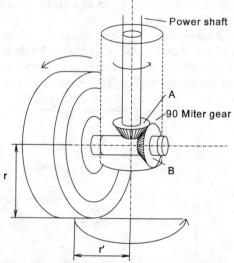

Figure 1.10: Slip compensation during a turn is accomplished through use of an offset foot assembly on the three-wheeled *K2A Navmaster* robot. (Adapted from [Holland, 1983].)

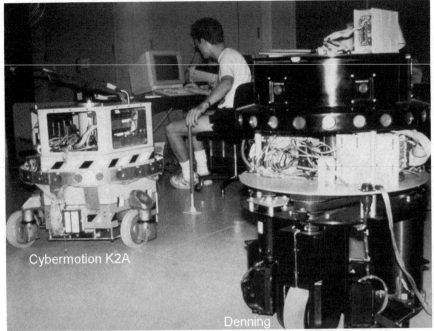

Figure 1.11: The Denning *Sentry* (foreground) incorporates a three-point *synchro-drive* configuration with each wheel located directly below the pivot axis of the associated steering column. In contrast, the Cybermotion *K2A* (background) has wheels that swivel around the steering column. Both robots were extensively tested at the University of Michigan's Mobile Robotics Lab. (Courtesy of The University of Michigan.)

$$D = \frac{2\pi N}{C_e} R_e \qquad\qquad\qquad (1.12)$$

where
D = vehicle displacement along path
N = measured counts of drive motor shaft encoder
C_e = encoder counts per complete wheel revolution
R_e = effective wheel radius.

1.3.5 Omnidirectional Drive

The odometry solution for most multi-degree-of-freedom (MDOF) configurations is done in similar fashion to that for differential drive, with position and velocity data derived from the motor (or wheel) shaft encoders. For the three-wheel example illustrated in Figure 1.12, the equations

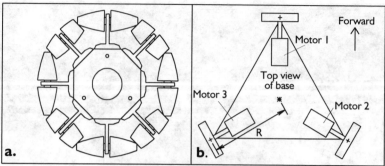

Figure 1.12: a. Schematic of the wheel assembly used by the Veterans Administration
[La et al., 1981] on an omnidirectional wheelchair.
b. Top view of base showing relative orientation of components in the
three-wheel configuration. (Adapted from [Holland, 1983].)

of motion relating individual motor speeds to velocity components V_x and V_y in the reference frame
of the vehicle are given by [Holland, 1983]:

$$V_1 = \omega_1 r = V_x + \omega_p R$$
$$V_2 = \omega_2 r = -0.5 V_x + 0.867 V_y + \omega_p R \qquad\qquad (1.13)$$
$$V_3 = \omega_3 r = -0.5 V_x - 0.867 V_y + \omega_p R$$

where
V_i = tangential velocity of wheel number i
ω_i = rotational speed of motor number i
ω_p = rate of base rotation about pivot axis
ω_r = effective wheel radius
ω_R = effective wheel offset from pivot axis.

1.3.6 Multi-Degree-of-Freedom Vehicles

Multi-degree-of-freedom (MDOF) vehicles have multiple
drive and steer motors. Different designs are possible. For
example, HERMIES-III, a sophisticated platform designed
and built at the Oak Ridge National Laboratory [Pin et al.,
1989; Reister et al., 1991; Reister, 1991] has two powered
wheels that are also individually steered (see Figure 1.13).
With four independent motors, HERMIES-III is a
4-degree-of-freedom vehicle.

MDOF configurations display exceptional maneuver-
ability in tight quarters in comparison to conventional
2-DOF mobility systems, but have been found to be
difficult to control due to their overconstrained nature

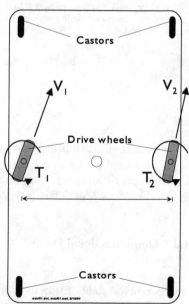

Figure 1.13: A 4-degree-of-freedom
vehicle platform can travel in all directions,
including sideways and diagonally. The
difficulty lies in coordinating all four motors
so as to avoid slippage.

[Reister et al., 1991; Killough and Pin, 1992; Pin and Killough, 1994; Borenstein, 1995]. Resulting problems include increased wheel slippage and thus reduced odometry accuracy. Recently, Reister and Unseren [1992; 1993] introduced a new control algorithm based on *Force Control*. The researchers reported on a substantial reduction in wheel slippage for their two-wheel drive/two-wheel steer platform, resulting in a reported 20-fold improvement of accuracy. However, the experiments on which these results were based avoided *simultaneous* steering and driving of the two steerable drive wheels. In this way, the critical problem of coordinating the control of all four motors *simultaneously and during transients* was completely avoided.

Unique Mobility, Inc. built an 8-DOF vehicle for the U.S. Navy under an SBIR grant (see Figure 1.14). In personal correspondence, engineers from that company mentioned to us difficulties in controlling and coordinating all eight motors.

Figure 1.14: An 8-DOF platform with four wheels individually driven and steered. This platform was designed and built by *Unique Mobility, Inc.* (Courtesy of [UNIQUE].)

1.3.7 MDOF Vehicle with Compliant Linkage

To overcome the problems of control and the resulting excessive wheel slippage described above, researchers at the University of Michigan designed the unique *Multi-Degree-of-Freedom* (MDOF) vehicle shown in Figures 1.15 and 1.16 [Borenstein, 1992; 1993; 1994c; 1995]. This vehicle comprises two differential-drive *LabMate* robots from [TRC]. The two *LabMates*, here referred to as "trucks," are connected by a *compliant linkage* and two rotary joints, for a total of three internal degrees of freedom.

The purpose of the compliant linkage is to accommodate momentary controller errors without transferring any mutual force reactions between the trucks, thereby eliminating the excessive wheel slippage reported for other MDOF vehicles. Because it eliminates excessive wheel slippage, the MDOF vehicle with compliant linkage is one to two orders of magnitude more accurate than other MDOF vehicles, and as accurate as conventional, 2-DOF vehicles.

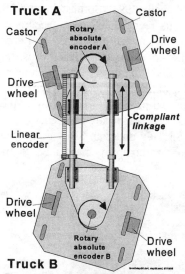

Figure 1.15: The compliant linkage is instrumented with two absolute rotary encoders and a linear encoder to measure the relative orientations and separation distance between the two trucks.

Figure 1.16: The University of Michigan's MDOF vehicle is a dual-differential-drive multi-degree-of-freedom platform comprising two TRC *LabMates*. These two "trucks" are coupled together with a *compliant linkage*, designed to accommodate momentary controller errors that would cause excessive wheel slippage in other MDOF vehicles. (Courtesy of The University of Michigan.)

1.3.8 Tracked Vehicles

Yet another drive configuration for mobile robots uses tracks instead of wheels. This very special implementation of a differential drive is known as *skid steering* and is routinely implemented in track form on bulldozers and armored vehicles. Such skid-steer configurations intentionally rely on track or wheel slippage for normal operation (Figure 1.17), and as a consequence provide rather poor dead-reckoning information. For this reason, skid steering is gener-

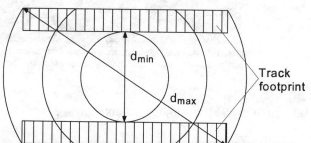

Figure 1.17: The effective point of contact for a skid-steer vehicle is roughly constrained on either side by a rectangular zone of ambiguity corresponding to the track footprint. As is implied by the concentric circles, considerable slippage must occur in order for the vehicle to turn [Everett, 1995].

ally employed only in tele-operated as opposed to autonomous robotic applications, where the ability to surmount significant floor discontinuities is more desirable than accurate odometry information. An example is seen in the track drives popular with remote-controlled robots intended for explosive ordnance disposal. Figure 1.18 shows the Remotec *Andros V* platform being converted to fully autonomous operation (see Sec. 5.3.1.2).

Figure 1.18: A Remotec *Andros V* tracked vehicle is outfitted with computer control at the University of Michigan. Tracked mobile platforms are commonly used in tele-operated applications. However, because of the lack of odometry feedback they are rarely (if at all) used in fully autonomous applications. (Courtesy of The University of Michigan.)

CHAPTER 2
HEADING SENSORS

Heading sensors are of particular importance to mobile robot positioning because they can help compensate for the foremost weakness of odometry: in an odometry-based positioning method, any small *momentary* orientation error will cause a *constantly* growing lateral position error. For this reason it would be of great benefit if orientation errors could be detected and corrected immediately. In this chapter we discuss gyroscopes and compasses, the two most widely employed sensors for determining the heading of a mobile robot (besides, of course, odometry). Gyroscopes can be classified into two broad categories: (a) mechanical gyroscopes and (b) optical gyroscopes.

2.1 Mechanical Gyroscopes

The mechanical gyroscope, a well-known and reliable rotation sensor based on the inertial properties of a rapidly spinning rotor, has been around since the early 1800s. The first known gyroscope was built in 1810 by G.C. Bohnenberger of Germany. In 1852, the French physicist Leon Foucault showed that a gyroscope could detect the rotation of the earth [Carter, 1966]. In the following sections we discuss the principle of operation of various gyroscopes.

Anyone who has ever ridden a bicycle has experienced (perhaps unknowingly) an interesting characteristic of the mechanical gyroscope known as *gyroscopic precession*. If the rider leans the bike over to the left around its own horizontal axis, the front wheel responds by turning left around the vertical axis. The effect is much more noticeable if the wheel is removed from the bike, and held by both ends of its axle while rapidly spinning. If the person holding the wheel attempts to yaw it left or right about the vertical axis, a surprisingly violent reaction will be felt as the axle instead twists about the horizontal roll axis. This is due to the angular momentum associated with a spinning flywheel, which displaces the applied force by 90 degrees in the direction of spin. The rate of precession Ω is proportional to the applied torque T [Fraden, 1993]:

$$T = I \omega \Omega \qquad\qquad (2.1)$$

where
T = applied input torque
I = rotational inertia of rotor
ω = rotor spin rate
Ω = rate of precession.

Gyroscopic precession is a key factor involved in the concept of operation for the *north-seeking gyrocompass*, as will be discussed later.

Friction in the support bearings, external influences, and small imbalances inherent in the construction of the rotor cause even the best mechanical gyros to drift with time. Typical systems employed in inertial navigation packages by the commercial airline industry may drift about $0.1°$ during a 6-hour flight [Martin, 1986].

2.1.1 Space-Stable Gyroscopes

The earth's rotational velocity at any given point on the globe can be broken into two components: one that acts around an imaginary vertical axis normal to the surface, and another that acts around an imaginary horizontal axis tangent to the surface. These two components are known as the *vertical earth rate* and the *horizontal earth rate*, respectively. At the North Pole, for example, the component acting around the local vertical axis (vertical earth rate) would be precisely equal to the rotation rate of the earth, or 15°/hr. The horizontal earth rate at the pole would be zero.

As the point of interest moves down a meridian toward the equator, the vertical earth rate at that particular location decreases proportionally to a value of zero at the equator. Meanwhile, the horizontal earth rate, (i.e., that component acting around a horizontal axis tangent to the earth's surface) increases from zero at the pole to a maximum value of 15°/hr at the equator.

There are two basic classes of rotational sensing gyros: 1) rate gyros, which provide a voltage or frequency output signal proportional to the turning rate, and 2) rate integrating gyros, which indicate the actual turn angle [Udd, 1991]. Unlike the magnetic compass, however, rate integrating gyros can only measure relative as opposed to absolute angular position, and must be initially referenced to a known orientation by some external means.

A typical gyroscope configuration is shown in Figure 2.1. The electrically driven *rotor* is suspended in a pair of precision low-friction bearings at either end of the rotor axle. The *rotor* bearings are in turn supported by a circular ring, known as the *inner gimbal ring*; this inner gimbal ring pivots on a second set of bearings that attach it to the *outer gimbal ring*. This pivoting action of the inner gimbal defines the horizontal axis of the gyro, which is perpendicular to the spin axis of the rotor as shown in Figure 2.1. The outer gimbal ring is attached to the instrument frame by a third set of bearings that define the vertical axis of the gyro. The vertical axis is perpendicular to both the horizontal axis and the spin axis.

Notice that if this configuration is oriented such that the spin axis points east-west, the horizontal axis is aligned with the north-south meridian. Since the gyro is space-stable (i.e., fixed

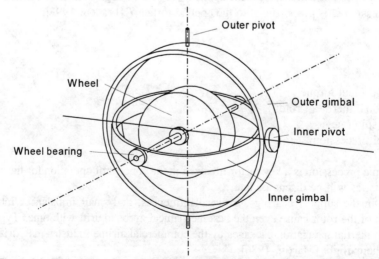

Figure 2.1: Typical two-axis mechanical gyroscope configuration [Everett, 1995].

in the inertial reference frame), the horizontal axis thus reads the horizontal earth rate component of the planet's rotation, while the vertical axis reads the vertical earth rate component. If the spin axis is rotated 90 degrees to a north-south alignment, the earth's rotation does not affect the gyro's horizontal axis, since that axis is now orthogonal to the horizontal earth rate component.

2.1.2 Gyrocompasses

The gyrocompass is a special configuration of the rate integrating gyroscope, employing a gravity reference to implement a north-seeking function that can be used as a true-north navigation reference. This phenomenon, first demonstrated in the early 1800s by Leon Foucault, was patented in Germany by Herman Anschutz-Kaempfe in 1903, and in the U.S. by Elmer Sperry in 1908 [Carter, 1966]. The U.S. and German navies had both introduced gyrocompasses into their fleets by 1911 [Martin, 1986].

The north-seeking capability of the gyrocompass is directly tied to the horizontal earth rate component measured by the horizontal axis. As mentioned earlier, when the gyro spin axis is oriented in a north-south direction, it is insensitive to the earth's rotation, and no tilting occurs. From this it follows that if tilting is observed, the spin axis is no longer aligned with the meridian. The direction and magnitude of the measured tilt are directly related to the direction and magnitude of the misalignment between the spin axis and true north.

2.1.3 Commercially Available Mechanical Gyroscopes

Numerous mechanical gyroscopes are available on the market. Typically, these precision machined gyros can cost between $10,000 and $100,000. Lower cost mechanical gyros are usually of lesser quality in terms of drift rate and accuracy. Mechanical gyroscopes are rapidly being replaced by modern high-precision — and recently — low-cost fiber-optic gyroscopes. For this reason we will discuss only a few low-cost mechanical gyros, specifically those that may appeal to mobile robotics hobbyists.

2.1.3.1 Futaba Model Helicopter Gyro

The Futaba FP-G154 [FUTABA] is a low-cost low-accuracy mechanical rate gyro designed for use in radio-controlled model helicopters and model airplanes. The Futaba FP-G154 costs less than $150 and is available at hobby stores, for example [TOWER]. The unit comprises of the mechanical gyroscope (shown in Figure 2.2 with the cover removed) and a small control amplifier. Designed for weight-sensitive

Figure 2.2: The Futaba FP-G154 miniature mechanical gyroscope for radio-controlled helicopters. The unit costs less than $150 and weighs only 102 g (3.6 oz).

model helicopters, the system weighs only 102 grams (3.6 oz). Motor and amplifier run off a 5 V DC supply and consume only 120 mA. However, sensitivity and accuracy are orders of magnitude lower than "professional" mechanical gyroscopes. The drift of radio-control type gyroscopes is on the order of tens of degrees per minute.

2.1.3.2 Gyration, Inc.

The *GyroEngine* made by Gyration, Inc. [GYRATION], Saratoga, CA, is a low-cost mechanical gyroscope that measures changes in rotation around two independent axes. One of the original applications for which the GyroEngine was designed is the *GyroPoint*, a three-dimensional pointing device for manipulating a cursor in three-dimensional computer graphics. The *GyroEngine* model GE9300-C has a typical drift rate of about 9°/min. It weighs only 40 grams (1.5 oz) and compares in size with that of a roll of 35 millimeter film (see Figure 2.3). The sensor can be

Figure 2.3: The Gyration *GyroEngine* compares in size favorably with a roll of 35 mm film (courtesy Gyration, Inc.).

powered with 5 to 15 VDC and draws only 65 to 85 mA during operation. The open collector outputs can be readily interfaced with digital circuits. A single *GyroEngine* unit costs $295.

2.2 Optical Gyroscopes

Optical rotation sensors have now been under development as replacements for mechanical gyros for over three decades. With little or no moving parts, such devices are virtually maintenance free and display no gravitational sensitivities, eliminating the need for gimbals. Fueled by a large anticipated market in the automotive industry, highly linear fiber-optic versions are now evolving that have wide dynamic range and very low projected costs.

The principle of operation of the optical gyroscope, first discussed by Sagnac [1913], is conceptually very simple, although several significant engineering challenges had to be overcome before practical application was possible. In fact, it was not until the demonstration of the helium-neon laser at Bell Labs in 1960 that Sagnac's discovery took on any serious implications; the first operational ring-laser gyro was developed by Warren Macek of Sperry Corporation just two years later [Martin, 1986]. Navigation quality ring-laser gyroscopes began routine service in inertial navigation systems for the Boeing 757 and 767 in the early 1980s, and over half a million fiber-optic navigation systems have been installed in Japanese automobiles since 1987 [Reunert, 1993]. Many technological improvements since Macek's first prototype make the optical rate gyro a potentially significant influence on mobile robot navigation in the future.

The basic device consists of two laser beams traveling in opposite directions (i.e., counter propagating) around a closed-loop path. The constructive and destructive interference patterns formed by splitting off and mixing parts of the two beams can be used to determine the rate and direction of rotation of the device itself.

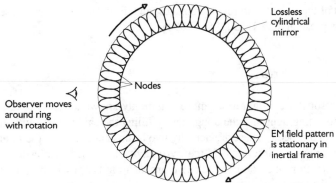

Figure 2.4: Standing wave created by counter-propagating light beams in an idealized ring-laser gyro. (Adapted from [Schulz-DuBois, 1966].)

Schulz-DuBois [1966] idealized the ring laser as a hollow doughnut-shaped mirror in which light follows a closed circular path. Assuming an ideal 100-percent reflective mirror surface, the optical energy inside the cavity is theoretically unaffected by any rotation of the mirror itself. The counter-propagating light beams mutually reinforce each other to create a stationary standing wave of intensity peaks and nulls as depicted in Figure 2.4, regardless of whether the gyro is rotating [Martin, 1986]. A simplistic visualization based on the Schulz-DuBois idealization is perhaps helpful at this point in understanding the fundamental concept of operation before more detailed treatment of the subject is presented. The light and dark fringes of the nodes are analogous to the reflective stripes or slotted holes in the rotating disk of an incremental optical encoder, and can be theoretically counted in similar fashion by a light detector mounted on the cavity wall. (In this analogy, however, the standing-wave "disk" is fixed in the inertial reference frame, while the normally stationary detector revolves around it.) With each full rotation of the mirrored doughnut, the detector would see a number of node peaks equal to twice the optical path length of the beams divided by the wavelength of the light.

Obviously, there is no practical way to implement this theoretical arrangement, since a *perfect* mirror cannot be realized in practice. Furthermore, the introduction of light energy into the cavity (as well as the need to observe and count the nodes on the standing wave) would interfere with the mirror's performance, should such an ideal capability even exist. However, many practical embodiments of optical rotation sensors have been developed for use as rate gyros in navigation applications. Five general configurations will be discussed in the following subsections:

- Active optical resonators (2.2.1).
- Passive optical resonators (2.2.2).
- Open-loop fiber-optic interferometers (analog) (2.2.3).
- Closed-loop fiber-optic interferometers (digital) (2.2.4).
- Fiber-optic resonators (2.2.5).

Aronowitz [1971], Menegozzi and Lamb [1973], Chow et al. [1985], Wilkinson [1987], and Udd [1991] provide in-depth discussions of the theory of the ring-laser gyro and its fiber-optic derivatives. A comprehensive treatment of the technologies and an extensive bibliography of preceding works is presented by Ezekial and Arditty [1982] in the proceedings of the First International Conference on Fiber-Optic Rotation Sensors held at MIT in November, 1981. An excellent treatment of the salient features, advantages, and disadvantages of ring laser gyros versus fiber optic gyros is presented by Udd [1985, 1991].

2.2.1 Active Ring Laser Gyros

The active optical resonator configuration, more commonly known as the ring laser gyro, solves the problem of introducing light into the doughnut by filling the cavity itself with an active *lazing* medium, typically helium-neon. There are actually two beams generated by the laser, which travel around the ring in opposite directions. If the gyro cavity is caused to physically rotate in the counterclockwise direction, the counterclockwise propagating beam will be forced to traverse a slightly longer path than under stationary conditions. Similarly, the clockwise propagating beam will see its closed-loop path shortened by an identical amount. This phenomenon, known as the *Sagnac effect*, in essence changes the length of the resonant cavity. The magnitude of this change is given by the following equation [Chow et al., 1985]:

$$\Delta L = \frac{4\pi r^2 \Omega}{c} \tag{2.2}$$

where
ΔL = change in path length
r = radius of the circular beam path
Ω = angular velocity of rotation
c = speed of light.

Note that the change in path length is directly proportional to the rotation rate Ω of the cavity. Thus, to measure gyro rotation, some convenient means must be established to measure the induced change in the optical path length.

This requirement to measure the difference in path lengths is where the invention of the laser in the early 1960s provided .the needed technological breakthrough that allowed Sagnac's observations to be put to practical use. For lazing to occur in the resonant cavity, the round-trip beam path must be precisely equal in length to an integral number of wavelengths at the resonant frequency. This means the wavelengths (and therefore the frequencies) of the two counter-propagating beams must change, as only oscillations with wavelengths satisfying the resonance condition can be sustained in the cavity. The frequency difference between the two beams is given by [Chow et al., 1985]:

$$\Delta f = \frac{2fr\Omega}{c} = \frac{2r\Omega}{\lambda} \tag{2.3}$$

where
Δf = frequency difference
r = radius of circular beam path
Ω = angular velocity of rotation
λ = wavelength.

In practice, a doughnut-shaped ring cavity would be hard to realize. For an arbitrary cavity geometry, the expression becomes [Chow et al., 1985]:

$$\Delta f = \frac{4A\Omega}{P\lambda}$$

(2.4)

where
Δf = frequency difference
A = area enclosed by the closed-loop beam path
Ω = angular velocity of rotation
P = perimeter of the beam path
λ = wavelength.

For single-axis gyros, the ring is generally formed by aligning three highly reflective mirrors to create a closed-loop triangular path as shown in Figure 2.5. (Some systems, such as Macek's early prototype, employ four mirrors to create a square path.) The mirrors are usually mounted to a monolithic glass-ceramic block with machined ports for the cavity bores and electrodes. Most modern three-axis units employ a square block cube with a total of six mirrors, each mounted to the center of a block face as shown in Figure 2.5. The most stable systems employ linearly polarized light and minimize circularly polarized components to avoid magnetic sensitivities [Martin, 1986].

The approximate quantum noise limit for the ring-laser gyro is due to spontaneous emission in the gain medium [Ezekiel and Arditty, 1982]. Yet, the ring-laser gyro represents the "best-case" scenario of the five general gyro configurations outlined above. For this reason the active ring-laser gyro offers the highest sensitivity and is perhaps the most accurate implementation to date.

The fundamental disadvantage associated with the active ring

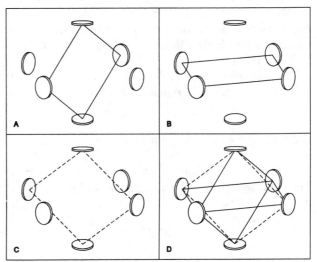

Figure 2.5: Six-mirror configuration of three-axis ring-laser gyro. (Adapted from [Koper, 1987].)

laser is a problem called *frequency lock-in*, which occurs at low rotation rates when the counter-propagating beams "lock" together in frequency [Chao et al., 1984]. This lock-in is attributed to the influence of a very small amount of backscatter from the mirror surfaces, and results in a deadband region (below a certain threshold of rotational velocity) for which there is no output signal. Above the lock-in threshold, output approaches the ideal linear response curve in a parabolic fashion.

The most obvious approach to solving the lock-in problem is to improve the quality of the mirrors to reduce the resulting backscatter. Again, however, perfect mirrors do not exist, and some finite amount of backscatter will always be present. Martin [1986] reports a representative value as 10^{-12} of the power of the main beam; enough to induce frequency lock-in for rotational rates of several hundred degrees per hour in a typical gyro with a 20-centimeter (8-in) perimeter.

An additional technique for reducing lock-in is to incorporate some type of biasing scheme to shift the operating point away from the deadband zone. Mechanical dithering is the least elegant but most common biasing means, introducing the obvious disadvantages of increased system complexity and reduced mean time between failures due to the moving parts. The entire gyro assembly is rotated back and forth about the sensing axis in an oscillatory fashion. State-of-the-art dithered active ring laser gyros have a scale factor linearity that far surpasses the best mechanical gyros.

Dithered biasing, unfortunately, is too slow for high-performance systems (i.e., flight control), resulting in oscillatory instabilities [Martin, 1986]. Furthermore, mechanical dithering can introduce crosstalk between axes on a multi-axis system, although some unibody three-axis gyros employ a common dither axis to eliminate this possibility [Martin, 1986].

Buholz and Chodorow [1967], Chesnoy [1989], and Christian and Rosker [1991] discuss the use of extremely short duration laser pulses (typically 1/15 of the resonator perimeter in length) to reduce the effects of frequency lock-in at low rotation rates. The basic idea is to reduce the cross-coupling between the two counter-propagating beams by limiting the regions in the cavity where the two pulses overlap. Wax and Chodorow [1972] report an improvement in performance of two orders of magnitude through the use of intracavity phase modulation. Other techniques based on non-linear optics have been proposed, including an approach by Litton that applies an external magnetic field to the cavity to create a directionally dependent phase shift for biasing [Martin, 1986].

Yet another solution to the lock-in problem is to remove the lazing medium from the ring altogether, effectively forming what is known as a passive ring resonator.

2.2.2 Passive Ring Resonator Gyros

The passive ring resonator gyro makes use of a laser source external to the ring cavity (Figure 2.6), and thus avoids the frequency lock-in problem which arises when the gain medium is internal to the cavity itself. The passive configuration also eliminates problems arising from changes in the optical path length within the interferometer due to variations in the index of refraction of the gain medium [Chow et al., 1985]. The theoretical quantum noise limit is determined by photon shot noise and is slightly higher (i.e., worse) than the theoretical limit seen for the active ring-laser gyro [Ezekiel and Arditty, 1982].

The fact that these devices use mirrored resonators patterned after their active ring predecessors means that their packaging is inherently bulky. However, fiber-optic technology now offers a low volume alternative. The fiber-optic derivatives also allow longer length multi-turn resonators, for increased sensitivity in smaller, rugged, and less expensive packages. As a consequence, the Resonant Fiber-Optic Gyro (RFOG), to be discussed in Section 2.1.2.5, has emerged as the most popular of the resonator configurations [Sanders, 1992].

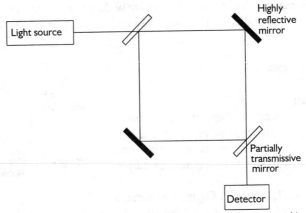

Figure 2.6: Passive ring resonator gyro with laser source external to the ring cavity. (Adapted from [Udd, 1991].)

2.2.3 Open-Loop Interferometric Fiber Optic Gyros

The concurrent development of optical fiber technology, spurred mainly by the communications industry, presented a potential low-cost alternative to the high-tolerance machining and clean-room assembly required for ring-laser gyros. The glass fiber in essence forms an internally reflective waveguide for optical energy, along the lines of a small-diameter linear implementation of the doughnut-shaped mirror cavity conceptualized by Schulz-DuBois [1966].

Recall the refractive index n relates the speed of light in a particular medium to the speed of light in a vacuum as follows:

$$n = \frac{c}{c_m} \tag{2.5}$$

where
n = refractive index of medium
c = speed of light in a vacuum
c_m = speed of light in medium.

Step-index multi-mode fiber (Figure 2.7) is made up of a core region of glass with index of refraction n_{co}, surrounded by a protective cladding with a lower index of refraction n_{cl} [Nolan and Blaszyk, 1991]. The lower refractive index in the cladding is necessary to ensure total internal reflection of the light propagating through the core region. The terminology *step index* refers to this "stepped" discontinuity in the refractive index that occurs at the core-cladding interface. Referring now to Figure 2.7, as long as the entry angle (with respect to the waveguide axis) of an incoming ray is less than a certain critical angle θ_c, the ray will be guided down the fiber, virtually without loss. The *numerical aperture* of the fiber quantifies this parameter of acceptance (the light-collecting ability of the fiber) and is defined as follows [Nolan and Blaszyk, 1991]:

$$NA = \sin\theta_c = \sqrt{n_{co}^2 - n_{cl}^2} \qquad (2.6)$$

where
NA = numerical aperture of the fiber
θ_c = critical angle of acceptance
n_{co} = index of refraction of glass core
n_{cl} = index of refraction of cladding.

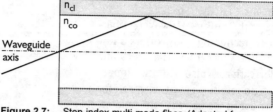

Figure 2.7:   Step-index multi-mode fiber. (Adapted from [Nolan et al., 1991].)

As illustrated in Figure 2.8, a number of rays following different-length paths can simultaneously propagate down the fiber, as long as their respective entry angles are less than the critical angle of acceptance θ_c. Multiple-path propagation of this nature occurs where the core diameter is much larger than the wavelength of the guided energy, giving rise to the term *multi-mode fiber*. Such multi-mode operation is clearly undesirable in gyro applications, where the objective is to eliminate all non-reciprocal conditions other than that imposed by the Sagnac effect itself. As the diameter of the core is reduced to approach the operating wavelength, a cutoff condition is reached where just a single mode is allowed to propagate, constrained to travel only along the waveguide axis [Nolan and Blaszyk, 1991].

Light can randomly change polarization states as it propagates through standard *single-mode fiber*. The use of special polarization-maintaining fiber, such as PRSM Corning, maintains the original polarization state of the light along the path of travel [Reunert, 1993]. This is important, since light of different polarization states travels through an optical fiber at different speeds.

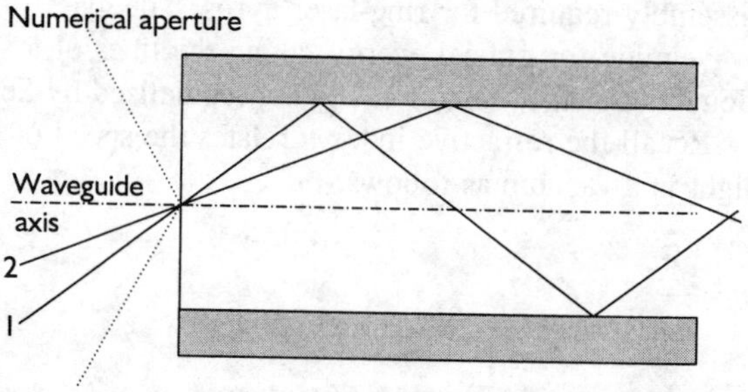

A typical block diagram of the "minimum-reciprocal" IFOG configuration is presented in Figure 2.9. Polarization-maintaining single-mode fiber

Figure 2.8: Entry angles of incoming rays 1 and 2 determine propagation paths in fiber core. (Adapted from [Nolan et al., 1991].)

[Nolan and Blaszyk, 1991] is employed to ensure the two counter-propagating beams in the loop follow identical paths in the absence of rotation.

An interesting characteristic of the IFOG is the absence of any laser source [Burns et al., 1983], the enabling technology allowing the Sagnac effect to reach practical implementation in the first place. A low-coherence source, such as a *super-luminescent diode* (SLD), is typically employed instead to reduce the effects of noise [Tai et al., 1986], the primary source of which is backscattering within the fiber and at any interfaces. As a result, in addition to the two primary counter-propagating waves in the loop, there are also a number of parasitic waves that yield

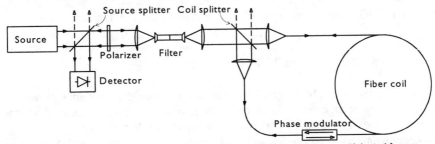

Figure 2.9: Block diagram of "minimum-reciprocal" integrated fiber-optic gyro. (Adapted from [Lefevre, 1992].)

secondary interferometers [Lefevre, 1992]. The limited temporal coherence of the broadband SLD causes any interference due to backscattering to average to zero, suppressing the contrast of these spurious interferometers. The detection system becomes sensitive only to the interference between waves that followed identical paths [Ezekiel and Arditty, 1982; Lefevre, 1992].

The Sagnac phase shift introduced by rotation is given by [Ezekiel and Arditty, 1982]

$$\Delta\phi = \frac{2\pi LD}{\lambda c} \tag{2.7}$$

where
$\Delta\phi$ = measured phase shift between counter-propagating beams
L = length of fiber-optic cable in loop
D = diameter of loop
λ = wavelength of optical energy
c = speed of light in a vacuum.

The stability of the scale factor relating $\Delta\phi$ to the rotational velocity in the equation above is thus limited to the stability of L, D, and λ [Ezekiel and Arditty, 1982]. Practical implementations usually operate over plus or minus half a fringe (i.e., $\pm\pi$ rad of phase difference), with a theoretical sensitivity of 10^{-6} radians or less of phase shift [Lefevre, 1992].

IFOG sensitivity may be improved by increasing L (i.e., adding turns of fiber in the sensing loop). This effect peaks at an optimal length of several kilometers, after which the fiber attenuation (typically 1 dB/km) begins to degrade performance. This large amount of fiber represents a significant percentage of overall system cost.

In summary, the open-loop IFOG is attractive from the standpoint of reduced manufacturing costs. Additional advantages include high tolerance to shock and vibration, insensitivity to gravity effects, quick start-up, and good sensitivity in terms of bias drift rate and the random walk coefficient. Coil geometry is not critical, and no path length control is needed. Some disadvantages are that a long optical cable is required, dynamic range is limited with respect to active ring-laser gyros, and the scale factor is prone to vary [Adrian, 1991]. Open-loop configurations are therefore most suited to the needs of low-cost systems in applications that require relatively low accuracy (i.e., automobile navigation).

For applications demanding higher accuracy, such as aircraft navigation (0.01 to 0.001°/hr), the closed-loop IFOG to be discussed in the next section offers significant promise.

2.2.4 Closed-Loop Interferometric Fiber Optic Gyros

This new implementation of a fiber-optic gyro provides feedback to a frequency or phase shifting element. The use of feedback results in the cancellation of the rotationally induced *Sagnac phase shift*. However, closed-loop digital signal processing is considerably more complex than the analog signal processing employed on open-loop IFOG configurations [Adrian, 1991]. Nonetheless, it now seems that the additional complexity is justified by the improved stability of the gyro: closed-loop IFOGs are now under development with drifts in the 0.001 to 0.01°/hr range, and scale-factor stabilities greater than 100 ppm (parts per million) [Adrian, 1991].

2.2.5 Resonant Fiber Optic Gyros

The *resonant fiber optic gyro* (RFOG) evolved as a solid-state derivative of the passive ring resonator gyro discussed in Section 2.1.2.2. In the solid-state implementation, a passive resonant cavity is formed from a multi-turn closed loop of optical fiber. An input coupler provides a means for injecting frequency-modulated light from a laser source into the resonant loop in both the clockwise and counterclockwise directions. As the frequency of the modulated light passes through a value such that the perimeter of the loop precisely matches an integral number of wavelengths at that frequency, input energy is strongly coupled into the loop [Sanders, 1992]. In the absence of loop rotation, maximum coupling for both beam directions occurs in a sharp peak centered at this resonant frequency.

If the loop is caused to rotate in the clockwise direction, of course, the *Sagnac effect* causes the perceived loop perimeter to lengthen for the clockwise-traveling beam, and to shorten for the counterclockwise-traveling beam. The resonant frequencies must shift accordingly, and as a result, energy is coupled into the loop at two different frequencies and directions during each cycle of the sinusoidal FM sweep. An output coupler samples the intensity of the energy in the loop by passing a percentage of the two counter-rotating beams to their respective detectors. The demodulated output from these detectors will show resonance peaks, separated by a frequency difference *f* given by the following [Sanders, 1992]:

$$\Delta f = \frac{D}{\lambda n} \Omega \tag{2.7}$$

where
Δf = frequency difference between counter-propagating beams
D = diameter of the resonant loop
Ω = rotational velocity
λ = freespace wavelength of laser
n = refractive index of the fiber.

Like the IFOG, the all-solid-state RFOG is attractive from the standpoint of high reliability, long life, quick start-up, and light weight. The principle advantage of the RFOG, however, is that it requires significantly less fiber (from 10 to 100 times less) in the sensing coil than the IFOG configuration, while achieving the same shot-noise-limited performance [Sanders, 1992]. Sanders attributes this to the fact that light traverses the sensing loop multiple times, as opposed to once in the IFOG counterpart. On the down side are the requirements for a highly coherent source and extremely low-loss fiber components [Adrian, 1991].

2.2.6 Commercially Available Optical Gyroscopes

Only recently have optical fiber gyros become commercially available at a price that is suitable for mobile robot applications. In this section we introduce two such systems.

2.2.6.1 The Andrew "Autogyro"

Andrew Corp. [ANDREW] offers the low-cost *Autogyro*, shown in Figure 2.10, for terrestrial navigation. It is a single-axis interferometric fiber-optic gyroscope (see Sec. 2.1.2.3) based on polarization-maintaining fiber and precision fiber-optic gyroscope technology. Model 3ARG-A ($950) comes with an analog output, while model 3ARG-D ($1,100) has an RS-232 output for connection to a computer. Technical specifications for the 3ARG-D are given in Table 2.1. Specifications for the 3ARG-A are similar. A more detailed discussion of the *Autogyro* is given in [Allen et al., 1994; Bennett and Emge, 1994].

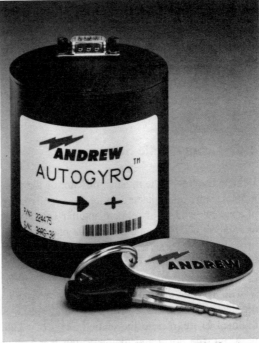

Table 2.1: Selected specifications for the Andrew *Autogyro* Model 3ARG-D. (Courtesy of [Andrew Corp].)

Parameter	Value	Units
Input rotation rate	±100	°/s
Minimum detectable rotation rate	±0.05	°/s
	±180	°/hr
Rate bandwidth	100	Hz
Bias drift (at stabilized temperature) — RMS	0.005	°/s rms
	18	°/hr rms
Size (excluding connector)	77 dia × 88	mm
	3.0 dia × 3.5	in
Weight (total)	0.63	kg
	1.38	lb
Power	9 to 18	VDC
	630	mA

Figure 2.10: The Andrew *Autogyro* Model 3ARG. (Courtesy of [Andrew Corp].)

2.2.6.2 Hitachi Cable Ltd. OFG-3

Hitachi Cable Ltd. markets an optical fiber gyroscope called OFG-3 (see Figure 2.11). Komoriya and Oyama [1994] tested that sensor and found its drift rate to be quite linear with 0.00317°/s (11.4°/hr). This result is close to the advertised specification of 10°/hr. This low drift rate is substantially better than that provided by conventional (mechanical) gyros. Table 2.2 shows technical specifications of the OFG-3 gyro, as reported by Komoriya and Oyama [1994].

One point to keep in mind when considering the use of fiber optic gyros in mobile robot applications is the minimum detectable rotation rate. This rate happens to be the same for both the Andrew 3ARG-A and the Hitachi OFG-3 gyros: 0.05°/s. If either gyro was installed on a robot with a systematic error (e.g., due to unequal wheel diameters; see Sec. 5.1 for more details) of 1 degree per 10 meter linear travel, then neither gyro would detect this systematic error at speeds lower than 0.5 m/s.

Table 2.2: Selected specifications for the Hitachi Cable Ltd. OFG-3 fiber optic gyroscope. (Reprinted with permission from [Komoriya and Oyama, 1994].)

Parameter	Value	Units
Input rotation rate	±100	°/s
Minimum detectable rotation rate	±0.05	°/s
	±60	°/hr
Min. sampl. interval	10	ms
Zero drift (rate integration)	0.0028	°/s
	10	°/hr
Size	88(W)×88(L)×65(H)	mm
	3.5(W)×3.5(L)×2.5(H)	in
Weight (total)	0.48	kg
	1.09	lb
Power	12	VDC
	150-250	mA

Figure 2.11: The OFG-3 optical fiber gyro made by Hitachi Cable Ltd. (Courtesy of Hitachi Cable America, Inc. [HITACHI].)

2.3 Geomagnetic Sensors

Vehicle heading is the most significant of the navigation parameters (x, y, and θ) in terms of its influence on accumulated dead-reckoning errors. For this reason, sensors which provide a measure of absolute heading or relative angular velocity are extremely important in solving the real world navigation needs of an autonomous platform. The most commonly known sensor of this type is probably the magnetic compass. The terminology normally used to describe the intensity of a magnetic field is *magnetic flux density B*, measured in Gauss (G). Alternative units are the Tesla (T), and the gamma (γ), where 1 Tesla = 10^4 Gauss = 10^9 gamma.

The average strength of the earth's magnetic field is 0.5 Gauss and can be represented as a dipole that fluctuates both in time and space, situated roughly 440 kilometers off center and inclined 11 degrees to the planet's axis of rotation [Fraden, 1993]. This difference in location between true north and magnetic north is known as *declination* and varies with both time and geographical location. Corrective values are routinely provided in the form of declination tables printed directly on the maps or charts for any given locale.

Instruments which measure magnetic fields are known as *magnetometers*. For application to mobile robot navigation, only those classes of magnetometers which sense the magnetic field of the earth are of interest. Such geomagnetic sensors, for purposes of this discussion, will be broken down into the following general categories:

- Mechanical magnetic compasses.
- Fluxgate compasses.
- Hall-effect compasses.
- Magnetoresistive compasses.
- Magnetoelastic compasses.

Before we introduce different types of compasses, a word of warning: the earth's magnetic field is often distorted near power lines or steel structures [Byrne et al., 1992]. This makes the straightforward use of geomagnetic sensors difficult for indoor applications. However, it may be possible to overcome this problem in the future by fusing data from geomagnetic compasses with data from other sensors.

2.3.1 Mechanical Magnetic Compasses

The first recorded use of a magnetic compass was in 2634 B.C., when the Chinese suspended a piece of naturally occurring magnetite from a silk thread and used it to guide a chariot over land [Carter, 1966]. Much controversy surrounds the debate over whether the Chinese or the Europeans first adapted the compass for marine applications, but by the middle of the 13th century such usage was fairly widespread around the globe. William Gilbert [1600] was the first to propose that the earth itself was the source of the mysterious magnetic field that provided such a stable navigation reference for ships at sea.

The early marine compasses were little more that magnetized needles floated in water on small pieces of cork. These primitive devices evolved over the years into the reliable and time proven systems in use today, which consist of a ring magnet or pair of bar magnets attached to a graduated mica readout disk. The magnet and disk assembly floats in a mixture of water and alcohol or glycerine, such that it is free to rotate around a jeweled pivot. The fluid acts to both support the weight of the rotating assembly and to dampen its motion under rough conditions.

The sealed vessel containing the compass disk and damping fluid is typically suspended from a 2-degree-of-freedom *gimbal* to decouple it from the ship's motion. This gimbal assembly is mounted in turn atop a floor stand or *binnacle*. On either side of the binnacle are massive iron spheres that, along with adjustable permanent magnets in the base, are used to compensate the compass for surrounding magnetic abnormalities that alter the geomagnetic lines of flux. The error resulting from such external influences (i.e., the angle between indicated and actual bearing to magnetic north) is known as compass *deviation*, and along with local declination, must be added or subtracted as appropriate for true heading:

$$H_t \ = \ H_i \ \pm \ CF_{\text{dev}} \ \pm \ CF_{\text{dec}} \tag{2.9}$$

where
H_t = true heading
H_i = indicated heading
CF_{dev} = correction factor for compass deviation
CF_{dec} = correction factor for magnetic declination.

Another potential source of error which must be taken into account is *magnetic dip*, a term arising from the "dipping" action observed in compass needles attributed to the vertical component of the geomagnetic field. The dip effect varies with latitude, from no impact at the equator where the flux lines are horizontal, to maximum at the poles where the lines of force are entirely vertical. For this reason, many swing-needle instruments have small adjustable weights that can be moved radially to balance the needle for any given local area of operation. Marine compasses ensure alignment in the horizontal plane by floating the magnet assembly in an inert fluid.

Dinsmore *Starguide* Magnetic Compass
An extremely low-cost configuration of the mechanical magnetic compass suitable for robotic applications is seen in a product recently announced by the Dinsmore Instrument Company, Flint, MI. The heart of the *Starguide* compass is the Dinsmore model 1490 digital sensor [Dinsmore Instrument Company, 1991], which consists of a miniaturized permanent-magnet rotor mounted in low-friction jeweled bearings. The sensor is internally damped such that if momentarily displaced 90 degrees, it will return to the indicated direction in 2.5 seconds, with no overshoot.

Four Hall-effect switches corresponding to the cardinal headings (N, E, W, S) are arranged around the periphery of the rotor and activated by the south pole of the magnet as the rotor aligns itself with the earth's magnetic field. Intermediate headings (NE, NW, SE, SW) are indicated through simultaneous activation of the adjacent cardinal-heading switches. The Dinsmore *Starguide* is not a true Hall-effect compass (see Sec. 2.3.3), in that the Hall-effect devices are not directly sensing the geomagnetic field of the earth, but rather the angular position of a mechanical rotor.

The model 1490 digital sensor measures 12.5 millimeters (0.5 in) in diameter by 16 millimeters (0.63 in) high, and is available separately from Dinsmore for around $12. Current consumption is 30 mA, and the open-collector NPN outputs can sink 25 mA per channel. Grenoble [1990] presents a simple circuit for interfacing the device to eight indicator LEDs. An alternative analog sensor (model 1525) with a ratiometric sine-cosine output is also available for around $35. Both sensors may be subjected to unlimited magnetic flux without damage.

2.3.2 Fluxgate Compasses

There currently is no practical alternative to the popular fluxgate compass for portability and long missions [Fenn et al., 1992]. The term *fluxgate* is actually a trade name of Pioneer Bendix for the *saturable-core magnetometer*, derived from the gating action imposed by an AC-driven excitation coil that induces a time varying permeability in the sensor core. Before discussing the principle of operation, it is probably best to review briefly the subject of magnetic conductance, or *permeability*.

The permeability μ of a given material is a measure of how well it serves as a path for magnetic lines of force, relative to air, which has an assigned permeability of one. Some examples of high-permeability materials are listed in Table 2.3.

Permeability is the magnetic circuit analogy to electrical conductivity, and relates magnetic flux density to the magnetizing force as follows:

$$B = \mu H \qquad (2.10)$$

where

B = magnetic flux density
μ = permeability
H = magnetizing force.

Table 2.3: Permeability ranges for selected materials. Values vary with proportional make-up, heat treatment, and mechanical working of the material [Bolz and Tuve, 1979].

Material	Permeability μ
Supermalloy	100,000 - 1,000,000
Pure iron	25,000 - 300,000
Mumetal	20,000 - 100,000
Permalloy	2,500 - 25,000
Cast iron	100 - 600

Since the magnetic flux in a magnetic circuit is analogous to current I in an electrical circuit, it follows that magnetic flux density B is the parallel to electrical current density.

A graphical plot of the above equation is known as the *normal magnetizing curve*, or B-H curve, and the permeability μ is the slope. An example plot is depicted in Figure 2.12 for the case of mild steel. In actuality, due to hysteresis, μ depends not only on the current value of H, but also the history of previous values and the sign of dH/dt, as will be seen later. The important thing to note at this point in the discussion is the *B-H* curve is not linear, but rather starts off with a fairly steep slope, and then flattens out suddenly as H reaches a certain value. Increasing H beyond this "knee" of the *B-H* curve yields little increase in B; the material is effectively saturated, with a near-zero permeability.

When a highly permeable material is introduced into a uniform magnetic field, the lines of force are drawn into the lower resistance path presented by the material as shown in Figure 2.13. However, if the material is forced into saturation by some additional magnetizing force H, the lines of flux of the external field will be relatively unaffected by the presence of the saturated material, as indicated in Figure 2.13b. The fluxgate magnetometer makes use of this saturation phenomenon in order to directly measure the strength of a surrounding static magnetic field.

Various core materials have been employed in different fluxgate designs over the past 50 years, with the two most common being permalloy (an alloy of iron and nickel) and mumetal (iron, nickel, copper, and chromium). The permeable core is driven into and out of saturation by a gating signal applied to an excitation coil wound around the core. For purposes

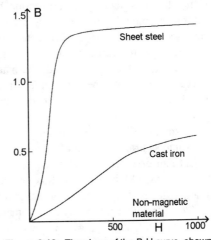

Figure 2.12: The slope of the B-H curve, shown here for cast iron and sheet steel, describes the permeability of a magnetic material, a measure of its ability (relative to air) to conduct a magnetic flux. (Adapted from [Carlson and Gisser, 1981].)

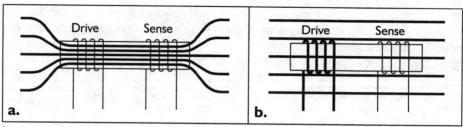

Figure 2.13: External lines of flux for: a. unsaturated core, b. saturated core. (Adapted from [Lenz, 1990].)

of illustration, let's assume for the moment a square-wave drive current is applied. As the core moves in and out of saturation, the flux lines from the external B field to be measured are drawn into and out of the core, alternating in turn between the two states depicted in Figure 2.13. (This is somewhat of an oversimplification, in that the *B-H* curve does not fully flatten out with zero slope after the knee.)

These expanding and collapsing flux lines will induce positive and negative EMF surges in a sensing coil properly oriented around the core. The magnitude of these surges will vary with the strength of the external magnetic field, and its orientation with respect to the axis of the core and sensing coil of the fluxgate configuration. The fact that the permeability of the sensor core can be altered in a controlled fashion by the excitation coil is the underlying principle which enables the DC field being measured to induce a voltage in the sense coil. The greater the differential between the saturated and unsaturated states (i.e., the steeper the slope), the more sensitive the instrument will be.

An idealized *B-H* curve for an alternating H-field is shown in Figure 2.14. The permeability (i.e., slope) is high along the section b-c of the curve, and falls to zero on either side of the saturation points H_s and $-H_s$, along segments c-d and a-b, respectively. Figure 2.14 shows a more representative situation: the difference between the left- and right-hand traces is due to hysteresis caused by some finite amount of permanent magnetization of the material. When a positive magnetizing force H_s is applied, the material will saturate with flux density B_s at point P_1 on the curve. When the magnetizing force is removed (i.e., $H = 0$), the flux density drops accordingly, but does not return to zero. Instead, there remains some residual magnetic flux density B_r, shown at point P_2, known as the retentivity.

A similar effect is seen in the application of an *H*-field of opposite polarity. The flux density goes into saturation at point P_3, then passes through point P_4 as the field reverses. This hysteresis effect can create what is known as a *zero offset* (i.e., some DC bias is still present when the external *B*-field is zero) in fluxgate magnetometers. Primdahl (1970) provides an excellent mathematical analysis of the actual gating curves for fluxgate devices.

The *effective permeability* μ_a of a material is influenced to a significant extent by its geometry. Bozorth and Chapin [1942] showed how μ_a for a cylindrical rod falls off with a decrease in the length-to-diameter ratio. This relationship can be attributed to the so-called *demagnetization factor* [Hine, 1968]. When a ferrous rod is coaxially aligned with the lines of flux of a magnetic field, a magnetic dipole is developed in the rod itself. The associated field introduced by the north and south poles of this dipole opposes the ambient field, with a corresponding reduction of flux density

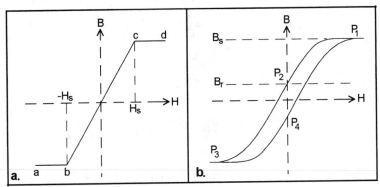

Figure 2.14: a. Ideal B-H curve.
b. Some minor hysteresis in the actual curve results in a residual non-zero value of *B*
when *H* is reduced to zero, known as the retentivity. (Adapted from Halliday and
Resnick, 1974; Carlson and Gisser, 1981).

through the rod. The lowered value of μ_a results in a less sensitive magnetometer, in that the
"flux-gathering" capability of the core is substantially reduced.

Consider again the cylindrical rod sensor presented in Figure 2.15, now in the absence of any
external magnetic field B_e. When the drive coil is energized, there will be a strong coupling
between the drive coil and the sense coil. Obviously, this will be an undesirable situation since the
output signal is supposed to be related to the strength of the external field only.

One way around this problem is seen in the Vacquier configuration developed in the early
1940s, where two parallel rods collectively form the core, with a common sense coil [Primdahl,
1979] as illustrated in Figure 2.15. The two rods are simultaneously forced into and out of
saturation, excited in antiphase by identical but oppositely wound solenoidal drive windings. In
this fashion, the magnetization fluxes of the two drive windings effectively cancel each other, with
no net effect on the sense coil.

Bridges of magnetic material may be employed
to couple the ends of the two coils together in a
closed-loop fashion for more complete flux linkage
through the core. This configuration is functionally
very similar to the ring-core design first employed
in 1928 by Aschenbrenner and Goubau [Geyger,
1957]. An alternative technique for decoupling the
pickup coil from the drive coil is to arrange the two
in an orthogonal fashion. In practice, there are a
number of different implementations of various
types of sensor cores and coil configurations as
described by Stuart [1972] and Primdahl [1979].
These are generally divided into two classes,
parallel and orthogonal, depending on whether the
excitation H-field is parallel or perpendicular to the
external B-field being measured. Alternative

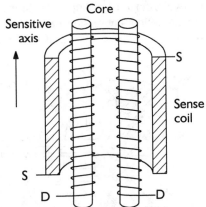

Figure 2.15: Identical but oppositely wound drive
windings in the Vacquier configuration cancel the
net effect of drive coupling into the surrounding
sense coil, while still saturating the core material.
(Adapted from [Primdahl, 1979].)

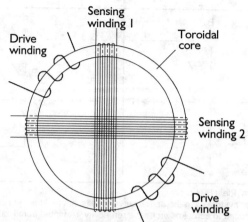

Figure 2.16: Two channel ring core fluxgate with toroidal excitation. (Adapted from [Acuna and Pellerin, 1969].)

excitation strategies (sine wave, square wave, sawtooth ramp) also contribute to the variety of implementations seen in the literature. Hine [1968] outlines four different classifications of saturable inductor magnetometers based on the method of readout (i.e., how the output EMF is isolated for evaluation):

- Fundamental frequency.
- Second harmonic.
- Peak output.
- Pulse difference.

Unambiguous 360-degree resolution of the earth's geomagnetic field requires two sensing coils at right angles to each other. The ring-core geometry lends itself to such dual-axis applications in that two orthogonal pickup coils can be configured in a symmetrical fashion around a common core. A follow-up version developed by Gordon and Lundsten [1970] employed a toroidal excitation winding as shown in Figure 2.17. Since there are no distinct poles in a closed-ring

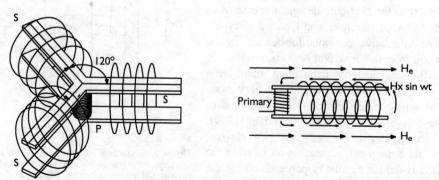

Figure 2.17: The Sperry *Flux Valve* consisted of a common drive winding P in the center of three sense windings S symmetrically arranged 120° apart. (Adapted from [Hine, 1968].)

design, demagnetization effects, although still present [Stuart, 1972], are less severe. The use of a ring geometry also leads to more complete flux linkage throughout the core, implying less required drive excitation for lower power operation, and the zero offset can be minimized by rotating the circular core. For these reasons, along with ease of manufacture, toroidal ring-core sensors are commonly employed in many of the low-cost fluxgate compasses available today.

The integrated DC output voltages V_x and V_y of the orthogonal sensing coils vary as sine and cosine functions of θ, where θ is the angle of the sensor unit relative to the earth's magnetic field. The instantaneous value of θ can be easily derived by performing two successive A/D conversions on these voltages and taking the arctangent of their quotient:

$$\theta = \arctan \frac{V_x}{V_y} . \tag{2.11}$$

Another popular two-axis core design is seen in the *Flux Valve* magnetometer developed by Sperry Corp. [SPERRY] and shown in Figure 2.17. This three-legged spider configuration employs three horizontal sense coils 120 degrees apart, with a common vertical excitation coil in the middle [Hine, 1968]. Referring to Figure 2.18, the upper and lower "arms" of the sense coil S are excited by the driving coil D, so that a magnetizing force H_x developed as indicated by the arrows. In the absence of an external field H_e, the flux generated in the upper and lower arms by the excitation coil is equal and opposite due to symmetry.

When this assembly is placed in an axial magnetic field H_e, however, the instantaneous excitation field H_x complements the flow in one arm, while opposing the flow in the other. This condition is periodically reversed in the arms, of course, due to the alternating nature of the driving function. A second-harmonic output is induced in the sensing coil S, proportional to the strength and orientation of the ambient field. By observing the relationships between the magnitudes of the output signals from each of the three sense coils (see Figure 2.18), the angular relationship of the *Flux Valve* with respect to the external field can be unambiguously determined.

When maintained in a level attitude, the fluxgate compass will measure the horizontal component of the earth's magnetic field, with the decided advantages of low power consumption, no moving parts, intolerance to shock ad vibration, rapid start-up, and relatively low cost. If the vehicle is expected to operate over uneven terrain, the sensor coil should be gimbal-mounted and mechanically dampened to prevent serious errors introduced by the vertical component of the geomagnetic field.

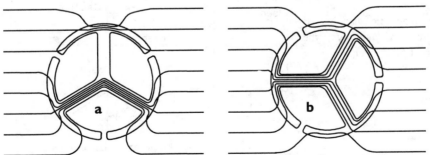

Figure 2.18: The *Flux Valve* magnetometer developed by Sperry Corporation uses a spider-core configuration. (Adapted from [Lenz, 1990].)

2.3.2.1 Zemco Fluxgate Compasses

The *Zemco fluxgate compass* [ZEMCO] was used in earlier work by Everett et al. [1990] on their robot called *ROBART II*. The sensor was a fluxgate compass manufactured by Zemco Electronics, San Ramon, CA, model number DE-700. This very low-cost (around $40) unit featured a rotating analog dial and was originally intended for 12 VDC operation in automobiles.

A system block diagram is presented in Figure 2.19. The sensor consists of two orthogonal pickup coils arranged around a toroidal excitation coil, driven in turn by a local oscillator. The outputs V_x and V_y of amplifier channels A and B are applied across an air-core resolver to drive the display indicator. The standard resolver equations [ILC Corporation, 1982] for these two voltages are

$$V_x = K_x \sin\theta \, \sin(\omega t + a_x) \tag{2.12a}$$
$$V_y = K_y \cos\theta \, \sin(\omega t + a_y) \tag{2.12b}$$

where
θ = the resolver shaft angle
ω = $2\pi f$, where f is the excitation frequency.
K_x and K_y are ideally equal transfer-function constants, and a_x and a_y are ideally zero time-phase shifts.

Thus, for any static spatial angle θ, the equations reduce to

$$V_x = K_x \sin\theta \tag{2.13a}$$
$$V_y = K_y \cos\theta \tag{2.13b}$$

which can be combined to yield

$$\frac{V_x}{V_Y} = \frac{\sin\theta}{\cos\theta} = \tan\theta \ . \tag{2.14}$$

The magnetic heading θ therefore is simply the arctangent of V_x over V_y.

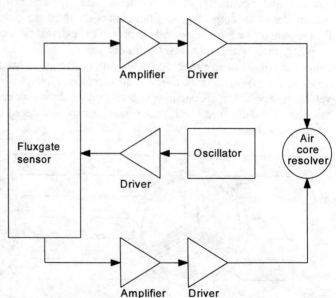

Figure 2.19: Block diagram of ZEMCO Model DE-700 fluxgate compass. (Courtesy of ZEMCO, Inc.)

Everett [1995] recounts his experience with two models of the Zemco fluxgate compass on *ROBART* II as follows:

Problems associated with the use of this particular fluxgate compass on ROBART, however, included a fairly high current consumption (250 mA), and stiction in the resolver reflecting back as a load into the drive circuitry, introducing some error for minor changes in vehicle heading. In addition, the sensor itself was affected by surrounding magnetic anomalies, some that existed on board the robot (i.e., current flow in nearby cable runs, drive and head positioning motors), and some present in the surrounding environment (metal desks, bookcases, large motors, etc.).

The most serious interference turned out to be the fluctuating magnetic fields due to power cables in close proximity — on the order of 30 centimeters (12 in) — to the fluxgate sensor. As various auxiliary systems on board the robot were turned on when needed and later deactivated to save power, the magnetic field surrounding the sensor would change accordingly. Serious errors could be introduced as well by minor changes in the position of cable runs, which occurred as a result of routine maintenance and trouble shooting. These problems were minimized by securing all cable runs with plastic tie-downs, and adopting a somewhat standardized protocol regarding which auxiliary systems would be activated when reading the compass.

There was no solution, however, for the interference effects of large metallic objects within the operating environment, and deviations of approximately four degrees were observed when passing within 30 centimeters (12 in) of a large metal cabinet, for example. A final source of error was introduced by virtue of the fact that the fluxgate compass had been mounted on the robot's head, so as to be as far away as possible from the effects of the drive motors and power distribution lines discussed above. The exact head position could only be read to within 0.82 degrees due to the limited resolution of the 8-bit A/D converter. In any event, an overall system error of ± 10 degrees was typical, and grossly insufficient for reliable dead-reckoning calculations, which was not the original intent of the compass.

This analog compass was later replaced by a newer digital version produced by Zemco, model DE-710, which cost approximately $90. The system block diagram is shown in Figure 2.20. This unit contained a built-in ADC0834 A/D converter to read the amplified outputs of the two sensor channels, and employed its own COP 421-MLA microprocessor, which drove a liquid crystal display (LCD). All communication between the A/D converter, microprocessor, and display driver was serial in nature, with a resulting slow update rate of 0.25 Hz. The built-in LCD simulated an analog dial with an extremely coarse resolution of $20°$ between

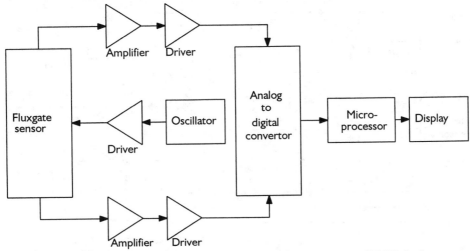

Figure 2.20: Block diagram of ZEMCO model DE-710 fluxgate compass (courtesy ZEMCO, Inc.).

display increments, but provision was made for serial output to an optional external shift register and associated three-digit numerical display.

All things considered, it was determined to be more practical to discard the built-in microprocessor, A/D converter, and LCD display, and interface an external A/D converter directly to the amplifier outputs as before with the analog version. This resulted in a decrease in supply current from 168 to 94 mA. Power consumption turned out to be less of a factor when it was discovered the circuitry could be powered up for a reading, and then deactivated afterwards with no noticeable effect on accuracy.

Overall system accuracy for this configuration was typically ±6 degrees, although a valid comparison to the analog version is not possible since the digital model was mounted in a different location to minimize interference from nearby circuitry. The amount of effort put into the calibration of the two systems must also be taken into account; the calibration procedure as performed was an iterative process not easily replicated from unit to unit with any quantitative measure.

2.3.2.2 Watson Gyrocompass

A combination fluxgate compass and solid-state rate gyro package (part number FGM-G100DHS-RS232) is available from Watson Industries, Eau Claire, WI [WATSON]. The system contains its own microprocessor that is intended to integrate the information from both the rate gyro and the compass to provide a more stable output less susceptible to interference, with an update rate of 40 Hz. An overall block diagram is presented in Figure 2.21.

The Watson fluxgate/rate gyro combination balances the shortcomings of each type of device: the gyro serves to filter out the effects of magnetic anomalies in the surrounding environment, while the compass counters the long-term drift of the gyro. Furthermore, the toroidal ring-core fluxgate sensor is gimbal-mounted for improved accuracy.

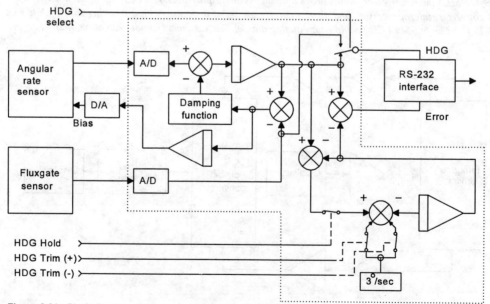

Figure 2.21: Block diagram of Watson fluxgate compass and rate gyro combination. (Courtesy of [WATSON].)

The Watson unit measures 6.3×4.4×7.6 centimeters (2.5×1.75×3.0 in) and weighs only 275 grams (10 oz). This integrated package is a much more expensive unit ($2,500) than the low-cost Zemco fluxgate compass, but is advertised to have higher accuracy (±2°). Power supply requirements are 12 VDC at 200 mA, and the unit provides an analog voltage output as well as a 12-bit digital output over a 2400-baud RS-232 serial link.

2.3.2.3 KVH Fluxgate Compasses

KVH Industries, Inc., Middletown, RI, offers a complete line of fluxgate compasses and related accessories, ranging from inexpensive units targeted for the individual consumer up through sophisticated systems intended for military applications [KVH]. The C100 COMPASS ENGINE (see Figure 2.22) is a versatile low-cost (less than $700) developer's kit that includes a microprocessor-controlled stand-alone fluxgate sensor subsystem based on a two-axis toroidal ring-core sensor.

Figure 2.22: The C-100 fluxgate compass engine was tested at the University of Michigan in a *flying robot* prototype. (Courtesy of [KVH].)

Two different sensor options are offered with the C-100: 1) the SE-25 sensor, recommended for applications with a tilt range of ±16 degrees and 2) the SE-10 sensor, for applications anticipating a tilt angle of up to ±45 degrees. The SE-25 sensor provides internal gimballing by floating the sensor coil in an inert fluid inside the lexan housing.The SE-10 sensor provides an additional 2-degree-of-freedom pendulous gimbal in addition to the internal fluid suspension. The SE-25 sensor mounts on top of the sensor PC board, while the SE-10 is suspended beneath it. The sensor PC board can be separated as much as 122 centimeters (48 in) from the detachable electronics PC board with an optional cable if so desired.

The resolution of the C100 is ±0.1 degrees, with an advertised accuracy of ±0.5 degrees (after compensation, with the sensor card level) and a repeatability of ±0.2 degrees. Separate ±180 degree adjustments are provided for *declination* as well as index offset (in the event the sensor unit cannot be mounted in perfect alignment with the vehicle's axis of travel). System damping can be user-selected, anywhere in the range of 0.1 to 24 seconds settling time to final value.

An innovative automatic compensation algorithm employed in the C100 is largely responsible for the high accuracy obtained by such a relatively low-priced system. This software routine runs on the controlling microprocessor mounted on the electronics board and corrects for magnetic anomalies associated with the host vehicle. Three alternative user-selectable procedures are offered:

- *Eight-Point Auto-Compensation* — starting from an arbitrary heading, the platform turns full circle, pausing momentarily at approximately 45-degree intervals. No known headings are required.
- *Circular Auto-Compensation* — Starting from an arbitrary position, the platform turns slowly through a continuous 360-degree circle. No known headings are required.
- *Three-Point Auto-Compensation* — Starting from an arbitrary heading, the platform turns and pauses on two additional known headings approximately 120 degrees apart.

Correction values are stored in a look-up table in non-volatile EEPROM memory. The automatic compensation routine also provides a quantitative indicator of the estimated quality of the current compensation and the magnitude of any magnetic interference present [KVH Industries, 1993].

The C100 configured with an SE-25 coil assembly weighs just 62 grams (2.25 oz) and draws 40 mA at 8 to 18 VDC (or 18 to 28 VDC). The combined sensor and electronics boards measure 4.6×11 centimeters (1.8×4.5 in). RS-232 (300 to 9600 baud) and NMEA 0183 digital outputs are provided, as well as linear and sine/cosine analog voltage outputs. Display and housing options are also available.

2.3.3 Hall-Effect Compasses

Hall-effect sensors are based on E. H. Hall's observation (in 1879) that a DC voltage develops across a conductor or semiconductor when in the presence of an external magnetic field. One advantage of this technology (i.e., relative to the fluxgate) is the inherent ability to directly sense a static flux, resulting in much simpler readout electronics. Early Hall magnetometers could not match the sensitivity and stability of the fluxgate [Primdahl, 1979], but the sensitivity of Hall devices has improved significantly. The more recent indium-antimonide devices have a lower sensitivity limit of 10^{-3} Gauss [Lenz, 1990].

The U.S. Navy in the early 1960s showed considerable interest in a small solid-state Hall-effect compass for low-power extended operations in sonobuoys [Wiley, 1964]. A number of such prototypes were built and delivered by Motorola for evaluation. The Motorola Hall-effect compass employed two orthogonal Hall elements for temperature-nulled non-ambiguous resolution of the geomagnetic field vector. Each sensor element was fabricated from a $2 \times 2 \times 0.1$ millimeter indium-arsenide-ferrite sandwich, and inserted between two wing-like mumetal flux concentrators as shown in Figure 2.23. It is estimated the 5 centimeter (2 in) magnetic concentrators increased the flux density through the sensing elements by two orders of magnitude [Wiley, 1964]. The output of the Motorola unit was a variable-width pulse train, the width of the pulse being proportional to the sensed magnetic heading. Excellent response linearity was reported down to flux densities of 0.001 Gauss [Willey, 1962].

Maenaka et al. [1990] report on the development of a monolithic silicon magnetic compass at the Toyohashi University of Technology in Japan, based on two orthogonal Hall-effect sensors. Their

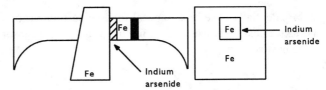

Figure 2.23: A pair of indium-arsenide-ferrite Hall-effect sensors (one shown) are positioned between flux concentrating wings of mumetal in this early Motorola prototype. (Adapted from [Wiley, 1964].)

use of the terminology "magnetic compass" is perhaps an unfortunate misnomer in that the prototype device was tested with an external field of 1,000 Gauss. Contrast this with the strength of the earth's magnetic field, which varies from only about 0.1 Gauss at the equator to about 0.9 Gauss at the poles. Silicon-based Hall-effect sensors have a lower sensitivity limit of around 10 Gauss [Lenz, 1990]. It is likely the Toyohashi University device was intended for other than geomagnetic applications, such as remote position sensing of rotating mechanical assemblies.

This prototype Hall-effect magnetometer is still of interest in that it represents a fully self-contained implementation of a two-axis magnetometer in integrated circuit form. Two vertical Hall cells [Maenaka et al., 1987] are arranged at right angles (see Figure 2.23) on a 4.7 mm² chip, with their respective outputs coupled to a companion signal processing IC of identical size. (Two separate chips were fabricated for the prototype instead of a single integrated unit to enhance production yield.) The sensor and signal processing ICs are interconnected (along with some external variable resistors for calibration purposes) on a glass-epoxy printed circuit board.

The dedicated signal-processing circuitry converts the B-field components B_x and B_y measured by the Hall sensors into an angle θ by means of the analog operation [Maenaka et al., 1990]:

$$\theta = \arctan\frac{B_x}{B_y} \tag{2.15}$$

where
θ = angle between B-field axis and sensor
B_x = x-component of B-field
B_y = y-component of B-field.

The analog output of the signal-processing IC is a DC voltage which varies linearly with vector orientation of the ambient magnetic field in a plane parallel to the chip surface. Reported test results show a fairly straight-line response (i.e., ± 2 percent full scale) for external field strengths ranging from 8,000 Gauss down to 500 Gauss; below this level performance begins to degrade rapidly [Maenaka et al., 1990]. A second analog output on the IC provides an indication of the absolute value of field intensity.

While the Toyohashi "magnetic compass" prototype based on silicon Hall-effect technology is incapable of detecting the earth's magnetic field, it is noteworthy nonetheless. A two-axis monolithic device of a similar nature employing the more sensitive indium-antimonide Hall devices could potentially have broad appeal for low-cost applications on mobile robotic platforms. An alternative possibility would be to use magnetoresistive sensor elements, which will be discussed in the next section.

2.3.4 Magnetoresistive Compasses

The general theory of operation for AMR and GMR magnetoresistive sensors for use in short-range proximity detection is beyond the scope of this text. However, there are three specific properties of the magnetoresistive magnetometer that make it well suited for use as a geomagnetic sensor: 1) high sensitivity, 2) directionality, and, in the case of AMR sensors, 3) the characteristic "flipping" action associated with the direction of internal magnetization.

AMR sensors have an open-loop sensitivity range of 10^{-2} Gauss to 50 Gauss (which easily covers the 0.1 to 1.0 Gauss range of the earth's horizontal magnetic field component), and limited-bandwidth closed-loop sensitivities approaching 10^{-6} Gauss [Lenz, 1990]. Excellent sensitivity, low power consumption, small package size, and decreasing cost make both AMR and GMR sensors increasingly popular alternatives to the more conventional fluxgate designs used in robotic vehicle applications.

2.3.4.1 Philips AMR Compass

One of the earliest magnetoresistive sensors to be applied to a magnetic compass application is the KMZ10B offered by Philips Semiconductors BV, The Netherlands [Dibburn and Petersen, 1983; Kwiatkowski and Tumanski, 1986; Petersen, 1989]. The limited sensitivity of this device (approximately 0.1 mV/A/m with a supply voltage of 5 VDC) in comparison to the earth's maximum horizontal magnetic field (15 A/m) means that considerable attention must be given to error-inducing effects of temperature and offset drift [Petersen, 1989].

One way around these problems is to exploit the "flipping" phenomenon by driving the device back and forth between its two possible magnetization states with square-wave excitation pulses applied to an external coil (Figure 2.24). This switching action toggles the sensor's axial magnetic field as shown in Figure 2.24a, resulting in the alternating response characteristics depicted in Figure 2.24b. Since the sensor offset remains unchanged while the signal output due to the external magnetic field H_y is inverted (Figure 2.24a), the undesirable DC offset voltages can be easily isolated from the weak AC signal.

A typical implementation of this strategy is shown in Figure 2.25. A 100 Hz square wave generator is capacitively coupled to the external excitation coil L which surrounds two orthogonally mounted magnetoresistive sensors. The sensors' output signals are amplified and AC-coupled to a synchronous detector driven by the same square-wave source. The rectified DC voltages V_{H1} and V_{H2} are thus proportional to the measured magnetic field components H_1 and H_2. The applied field direction is dependant on the ratio of V to H, not their absolute values. This means that as long as the two channels are calibrated to the same sensitivity, no temperature correction is required [Fraden, 1993].

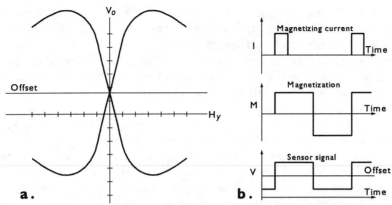

Figure 2.24: External current pulses set and reset the direction of magnetization, resulting in the "flipped" response characteristics shown by the dashed line. Note the DC offset of the device remains constant, while the signal output is inverted. (Adapted from [Petersen, 1989].)

2.3.5 Magnetoelastic Compasses

A number of researchers have recently investigated the use of *magnetoelastic* (also known as *magnetostrictive*) materials as sensing elements for high-resolution magnetometers. The principle of operation is based on the changes in Young's modulus experienced by magnetic alloys when exposed to an external magnetic field. The *modulus of elasticity E* of a given material is basically a measure of its stiffness, and directly relates stress to strain as follows:

$$E = \frac{\sigma}{\epsilon} \qquad (2.16)$$

where
E = Young's modulus of elasticity
σ = applied stress
ϵ = resulting strain.

Any ferromagnetic material will experience some finite amount of strain (expansion or shrinkage) in the direction of magnetization due to this *magnetostriction* phenomenon. It stands to reason that if the applied stress σ remains the same, strain ϵ will vary inversely with any change in Young's modulus E. In certain amorphous metallic alloys, this effect is very pronounced.

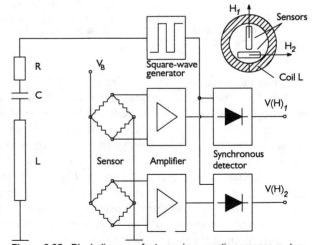

Figure 2.25: Block diagram of a two-axis magnetic compass system based on a commercially available anisotropic magnetoresistive sensor from Philips [Petersen, 1989].

Barrett et al. [1973] proposed a qualitative explanation, wherein individual atoms in the crystal lattice are treated as tiny magnetic dipoles. The forces exerted by these dipoles on one another depend upon their mutual orientation within the lattice; if the dipoles are aligned end to end, the opposite poles attract, and the material shrinks ever so slightly. The crystal is said to exhibit a negative magnetostriction constant in this direction. Conversely, if the dipoles are rotated into side-by-side alignment through the influence of some external field, like poles will repel, and the result is a small expansion.

It follows that the strength of an unknown magnetic field can be accurately measured if a suitable means is employed to quantify the resulting change in length of some appropriate material displaying a high magnetostriction constant. There are currently at least two measurement technologies with the required resolution allowing the magnetoelastic magnetometer to be a realistic contender for high-sensitivity low-cost performance: 1) *interferometric* displacement sensing, and 2) *tunneling-tip* displacement sensing.

Lenz [1990] describes a *magnetoelastic* magnetometer which employs a *Mach-Zender* fiber-optic interferometer to measure the change in length of a magnetostrictive material when exposed to an external magnetic field. A laser source directs a beam of light along two optical fiber paths by way of a beam splitter as shown in Figure 2.26. One of the fibers is coated with a material (nickel iron was used) exhibiting a high magnetostrictive constant. The length of this fiber is stretched or compressed in conjunction with any magnetoelastic expansion or contraction of its coating. The output beam from this fiber-optic cable is combined in a light coupler with the output beam from the uncoated reference fiber and fed to a pair of photodetectors.

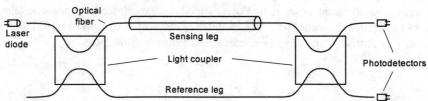

Figure 2.26: Fiber-optic magnetometers, basically a Mach-Zender interferometer with one fiber coated or attached to a magnetoelastic material, have a sensitivity range of 10^{-7} to 10 Gauss. (Adapted from [Lenz, 1990].)

Constructive and destructive interferences caused by differences in path lengths associated with the two fibers will cause the final output intensity as measured by the photodetectors to vary accordingly. This variation is directly related to the change in path length of the coated fiber, which in turn is a function of the magnetic field strength along the fiber axis. The prototype constructed by Lenz [1990] at Honeywell Corporation measured 10×2.5 centimeters (4×1 in) and was able to detect fields ranging from 10^7 Gauss up to 10 Gauss.

Researchers at the Naval Research Laboratory (NRL) have developed a prototype magnetoelastic magnetometer capable of detecting a field as small as 6×10^{-5} Gauss [Brizzolara et al., 1989] using the tunneling-tip approach. This new displacement sensing technology, invented in 1982 at IBM Zürich, is based on the measurement of current generated by quantum mechanical tunneling of electrons across a narrow gap (Figure 2.27). An analog feedback circuit compares the measured tunnel current with a desired value and outputs a drive signal to suitably adjust the distance between the tunneling electrodes with an electromechanical actuator [Kenny et al., 1991]. The instantaneous tunneling current is directly proportional to the exponential of electrode displacement. The most common actuators employed in this role are piezoelectric and electrostatic, the latter lending itself more readily to silicon micro-machining techniques.

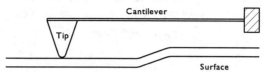

Figure 2.27: *Scanning tunneling microscopy,* invented at IBM Zürich in 1982, uses quantum mechanical tunneling of electrons across a barrier to measure separation distance at the gap. (Courtesy of T. W. Kenny, NASA JPL).

The active sense element in the NRL magnetometer is a 10 centimeter (4 in) metallic glass ribbon made from METGLAS 2605S2, annealed in a transverse magnetic field to yield a high magnetomechanical coupling [Brizzolara et al., 1989]. (METGLAS is an alloy of iron, boron, silicon, and carbon, and is a registered trademark of Allied Chemical.) The magnetoelastic ribbon elongates when exposed to an axial magnetic field, and the magnitude of this displacement is measured by a tunneling transducer as illustrated in Figure 2.28.

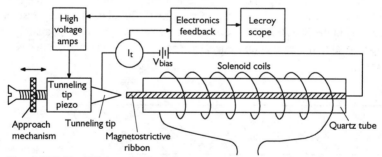

Figure 2.28: The NRL tunneling-transducer magnetometer employed a 10 cm (4 in) magnetoelastic ribbon vertically supported in a quartz tube [Brizzolara et al., 1989].

An electrochemically etched gold tip is mounted on a tubular piezoelectric actuator and positioned within about one nanometer of the free end of the METGLAS ribbon. The ribbon and tip are electrically biased with respect to each other, establishing a tunneling current that is fed back to the piezo actuator to maintain a constant gap separation. The degree of magnetically induced elongation of the ribbon can thus be inferred from the driving voltage applied to the piezoelectric actuator. The solenoidal coil shown in the diagram supplies a bias field of 0.85 oersted to shift the sensor into its region of maximum sensitivity.

Fenn et al. [1992] propose an alternative tunneling-tip magnetoelastic configuration with a predicted sensitivity of 2×10^{-11} Gauss, along the same order of magnitude as the cryogenically cooled SQUID. A small cantilevered beam of METGLAS 2605S2, excited at its resonant frequency by a gold-film electrostatic actuator, is centered between two high-permeability magnetic flux concentrators as illustrated in Figure 2.29. Any changes in the modulus of elasticity of the beam will directly affect its natural frequency; these changes in natural frequency can then be measured and directly related to the strength of the ambient magnetic field. The effective shift in natural frequency is rather small, however (Fenn et al. [1992] report only a 6 Hz shift at saturation), again necessitating a very precise method of measurement.

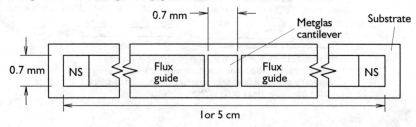

Figure 2.29: Top view of the single cantilevered design. (Adapted from [Fenn, et al., 1992].)

A second (non-magnetic) cantilever element is employed to track the displacement of the METGLAS reed with sub-angstrom resolution using tunneling-tip displacement sensing as illustrated in Figure 2.30. A pair of electrostatic actuator plates dynamically positions the reed follower to maintain a constant tunneling current in the probe gap, thus ensuring a constant lateral separation between the probe tip and the vibrating reed. The frequency of the excitation signal applied to the reed-follower actuator is therefore directly influenced by any resonant frequency changes occurring in the METGLAS reed. The magnetometer provides an analog voltage output which is proportional to this excitation frequency, and therefore indicative of the external magnetic field amplitude.

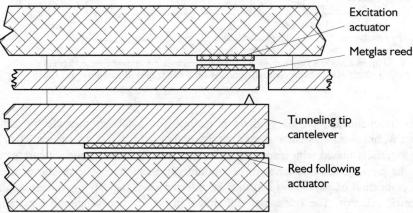

Figure 2.30: Side view of the double cantilevered design. (Adapted from [Fenn et al., 1992].)

CHAPTER 3
ACTIVE BEACONS

In this chapter we discuss sensors used for active beacon navigation. Active beacons have been used for many centuries as a reliable and accurate means for navigation. Stars can be considered as active beacons with respect to navigation; and lighthouses were early man-made beacon systems. Typical non-robotics applications for active beacon navigation include marine navigation, aircraft navigation, race car performance analysis, range instrumentation, unmanned mobile target control, mine localization, hazardous materials mapping, dredge positioning, geodetic surveys, and most recently, position location and range information for golfers [Purkey, 1994].

Modern technology has vastly enhanced the capabilities of active beacon systems with the introduction of laser, ultrasonic, and radio-frequency (RF) transmitters. Possibly the most promising approach to outdoor navigation is the recently completed *Global Positioning System* (GPS). This subject will be addressed in Section 3.1, while Section 3.2 introduces ground-based RF systems. It should be noted that according to our conversations with manufacturers, none of the RF systems can be used reliably in indoor environments. Non-RF beacon systems, based on ultrasonic and light sources, will be discussed in Sections 3.3 and 3.4.

3.1 Navstar Global Positioning System (GPS)

The recent Navstar Global Positioning System (GPS) developed as a Joint Services Program by the Department of Defense uses a constellation of 24 satellites (including three spares) orbiting the earth every 12 hours at a height of about 10,900 nautical miles. Four satellites are located in each of six planes inclined 55 degrees with respect to the plane of the earth's equator [Getting, 1993]. The absolute three-dimensional location of any GPS receiver is determined through simple trilateration techniques based on time of flight for uniquely coded spread-spectrum radio signals transmitted by the satellites. Precisely measured signal propagation times are converted to *pseudoranges* representing the line-of-sight distances between the receiver and a number of reference satellites in known orbital positions. The measured distances have to be adjusted for receiver clock offset, as will be discussed later, hence the term pseudoranges. Knowing the exact distance from the ground receiver to three satellites theoretically allows for calculation of receiver latitude, longitude, and altitude.

Although conceptually very simple (see [Hurn, 1993]), this design philosophy introduces at least four obvious technical challenges:
- Time synchronization between individual satellites and GPS receivers.
- Precise real-time location of satellite position.
- Accurate measurement of signal propagation time.
- Sufficient signal-to-noise ratio for reliable operation in the presence of interference and possible jamming.

The first of these problems is addressed through the use of atomic clocks (relying on the vibration period of the cesium atom as a time reference) on each of the satellites to generate time ticks at a frequency of 10.23 MHz. Each satellite transmits a periodic pseudo-random code on two different frequencies (designated L1 and L2) in the internationally assigned navigational frequency band. The L1 and L2 frequencies of 1575.42 and 1227.6 MHz are generated by multiplying the cesium-clock time ticks by 154 and 128, respectively. The individual satellite clocks are monitored by dedicated ground tracking stations operated by the Air Force, and continuously advised of their measured offsets from the ground master station clock. High precision in this regard is critical since electro-magnetic radiation propagates at the speed of light, roughly 0.3 meters (1 ft) per nanosecond.

To establish the exact time required for signal propagation, an identical pseudocode sequence is generated in the GPS receiver on the ground and compared to the received code from the satellite. The locally generated code is shifted in time during this comparison process until maximum correlation is observed, at which point the induced delay represents the time of arrival as measured by the receiver's clock. The problem then becomes establishing the relationship between the atomic clock on the satellite and the inexpensive quartz-crystal clock employed in the GPS receiver. This ΔT is found by measuring the range to a fourth satellite, resulting in four independent trilateration equations with four unknowns. Details of the mathematics involved are presented by Langley [1991].

The precise real-time location of satellite position is determined by a number of widely distributed tracking and telemetry stations at surveyed locations around the world. Referring to Figure 3.1, all measured and received data are forwarded to a master station for analysis and referenced to universal standard time. Change orders and signal-coding corrections are generated by the master station and then sent to the satellite control facilities for uploading [Getting, 1993]. In this fashion the satellites are continuously advised of their current position as perceived by the earth-based tracking stations, and encode this *ephemeris* information into their L1 and L2 transmissions to the GPS receivers. (Ephemeris is the space vehicle orbit characteristics, a set of numbers that precisely describe the vehicle's orbit when entered into a specific group of equations.)

In addition to its own timing offset and orbital information, each satellite transmits data on all other satellites in the constellation to enable any ground receiver to build up an almanac after a "cold start." Diagnostic information with respect to the status of certain onboard systems and expected range-measurement accuracy is also included. This collective "housekeeping" message is superimposed on the pseudo-random code modulation at a very low (50 bits/s) data rate, and requires 12.5 minutes for complete downloading [Ellowitz, 1992]. Timing offset and ephemeris information is repeated at 30 second intervals during this procedure to facilitate initial pseudorange measurements.

To further complicate matters, the sheer length of the unique pseudocode segment assigned to each individual Navstar Satellite (i.e., around 6.2 trillion bits) for repetitive transmission can potentially cause initial synchronization by the ground receiver to take considerable time. For this and other reasons, each satellite broadcasts two different non-interfering pseudocodes. The first of these is called the *coarse acquisition*, or C/A code, and is transmitted on the L1 frequency to assist in acquisition. There are 1023 different C/A codes, each having 1023 chips (code bits) repeated 1000 times a second [Getting, 1993] for an effective chip rate of 1.023 MHz (i.e., one-tenth the cesium clock rate). While the C/A code alone can be employed by civilian users to obtain a fix, the resultant positional accuracy is understandably somewhat degraded. The Y code (formerly the precision or P code prior to encryption on January 1st, 1994) is transmitted on both the L1 and L2 frequencies and scrambled for reception by authorized military users only with appropriate cryptographic keys and equipment. This encryption also ensures *bona fide* recipients cannot be "spoofed" (i.e., will not inadvertently track false GPS-like signals transmitted by unfriendly forces).

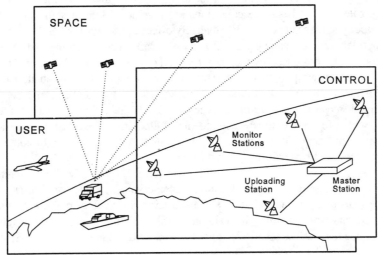

Figure 3.1: The Navstar *Global Positioning System* consists of three fundamental segments: Space, Control, and User. (Adapted from [Getting, 1993].)

Another major difference between the Y and C/A code is the length of the code segment. While the C/A code is 1023 bits long and repeats every millisecond, the Y code is 2.35×10^{14} bits long and requires 266 days to complete [Ellowitz, 1992]. Each satellite uses a one-week segment of this total code sequence; there are thus 37 unique Y codes (for up to 37 satellites) each consisting of 6.18×10^{12} code bits set to repeat at midnight on Saturday of each week. The higher chip rate of 10.23 MHz (equal to the cesium clock rate) in the precision Y code results in a chip wavelength of 30 meters for the Y code as compared to 300 meters for the C/A code [Ellowitz, 1992], and thus facilitates more precise time-of-arrival measurement for military purposes.

Brown and Hwang [1992] discuss a number of potential pseudorange error sources as summarized below in Table 3.1. Positional uncertainties related to the reference satellites are clearly a factor, introducing as much as 3 meters (9.8 ft) standard deviation in pseudo-range measurement accuracy. As the radiated signal propagates downward toward the earth, atmospheric refraction and multipath reflections (i.e., from clouds, land masses, water surfaces) can increase the perceived time of flight beyond that associated with the optimal straight-line path (Figure 3.2).

Table 3.1: Summary of potential error sources for measured pseudoranges [Brown and Hwang, 1992].

Error Source	Standard Deviation [m]	[ft]
Satellite position	3	29
Ionospheric refraction	5	16.4
Tropospheric refraction	2	6.6
Multipath reflection	5	16.4
Selective availability	30	98.4

Additional errors can be attributed to group delay uncertainties introduced by the processing and passage of the signal through the satellite electronics. Receiver noise and resolution must also be taken into account. Motazed [1993] reports fairly significant differences of 0.02 to 0.07 arc minutes in calculated latitudes and longitudes for two identical C/A-code receivers placed side by side. And finally, the particular dynamics of the mobile vehicle that hosts the GPS receiver plays a noteworthy role, in that best-case conditions are associated with a static platform, and any substantial velocity and acceleration will adversely affect the solution.

For commercial applications using the C/A code, small errors in timing and satellite position have been deliberately introduced by the master station to prevent a hostile nation from using GPS in support of precision weapons delivery. This intentional degradation in positional accuracy to around 100 meters (328 ft) best case and 200 meters (656 ft) typical *spherical error probable* (SEP) is termed *selective availability* [Gothard, 1993]. Selective availability has been on continuously (with a few exceptions) since the end of Operation Desert Storm. It was turned off during the war from August 1990 until July 1991 to improve the accuracy of commercial hand-held GPS receivers used by coalition ground forces.

There are two aspects of selective availability: *epsilon* and *dither*. Epsilon is intentional error in the navigation message regarding the location (ephemeris) of the satellite. *Dither* is error in the timing source (carrier frequency) that creates uncertainty in velocity measurements (Doppler). Some GPS receivers (for example, the Trimble ENSIGN) employ running-average filtering to statistically reduce the *epsilon* error over time to a reported value of 15 meters SEP [Wormley, 1994].

All of the error sources listed in Table 3.1 are further influenced by the particular geometry of the four reference satellites at time of sight-ing. Ignoring time synchronization needs for the moment (i.e., so only three satellites are required), the most accurate three-dimensional trilater-ation solutions will result when the bearing or sight lines extending from the receiver to the respective satellites

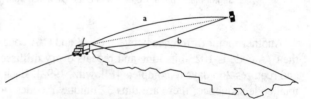

Figure 3.2: Contributing factors to pseudorange measurement errors: a. atmospheric refraction b. multi-path reflections [Everett, 1995].

are mutually orthogonal. If the satellites are spaced close together in a tight cluster or otherwise arranged in a more or less collinear fashion with respect to the receiver as shown in Figure 3.3, the desired orthogonality is lost and the solution degrades accordingly. This error multiplier, which can range from acceptable values of two or three all the way to infinity, is termed *geometric dilution of precision* [Byrne, 1993]. Kihara and Okada [1984] show where the minimum achievable (best-case) value for GDOP is 1.5811, and occurs when the four required GPS satellites are symmetrically located with an angle of 109.47 degrees between adjacent bearing lines as shown in Figure 3.4.

With the exception of multi-path effects, all of the error sources depicted in Table 3.1 above can be essentially eliminated through use of a practice known as *differential GPS* (DGPS). The concept is based on the premise that a second GPS receiver in fairly close proximity (i.e., within 10 km — 6.2 mi) to the first will experience basically the same error effects when viewing the same reference satellites. If this second receiver is fixed at a precisely surveyed location, its calculated solution can be compared to the known position to generate a composite error vector representative of prevailing conditions in that immediate locale. This differential correction can then be passed to the first receiver to null out the unwanted effects, effectively reducing position error for commercial systems to well under 10 meters.

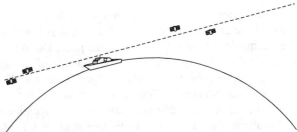

Figure 3.3: Worst-case *geometric dilution of precision* (GDOP) errors occur when the receiver and satellites approach a collinear configuration as shown [Everett, 1995].

The fixed DGPS reference station transmits these correction signals every two to four minutes to any differential-capable receiver within range. Many commercial GPS receivers are available with differential capability, and most now follow the RTCM-104 standard developed by the Radio Technical Commission for Maritime Services to promote interoperability. Prices for DGPS-capable mobile receivers run about $2K, while the reference stations cost somewhere between $10K and $20K. Magnavox is working with CUE Network Corporation to market a nationwide network to pass differential corrections over an FM link to paid subscribers [GPS Report, 1992].

Typical DGPS accuracies are around 4 to 6 meters (13 to 20 ft) SEP, with better performance seen as the distance between the mobile receivers and the fixed reference station is decreased. For example, the Coast Guard is in the process of implementing *differential GPS* in all major U.S. harbors, with an expected accuracy of around 1 meter (3.3 ft) SEP [Getting, 1993]. A differential GPS system already in operation at O'Hare International Airport in Chicago has demonstrated that aircraft and service vehicles can be located to 1 meter (3.3 ft). Surveyors use differential GPS to achieve centimeter accuracy, but this practice requires significant postprocessing of the collected data [Byrne, 1993].

An interesting variant of conventional DGPS is reported by Motazed [1993] in conjunction with the Non-Line-of-Sight Leader/Follower (NLOSLF) program underway at RedZone Robotics, Inc., Pittsburgh, PA. The NLOSLF

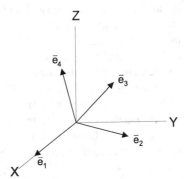

Figure 3.4: GDOP error contribution is minimal for four GPS satellites symmetrically situated with respect to the receiver (at origin) along bearing lines 109.47° apart [Kihara and Okada, 1984].

operational scenario involves a number of vehicles in a convoy configuration that autonomously follow a lead vehicle driven by a human operator, both on-road and off-road at varying speeds and separation distances. A technique to which Motazed refers as *intermittent stationary base differential GPS* is used to provide global referencing for purposes of bounding the errors of a sophisticated Kalman-filter-based GPS/INS position estimation system.

Under this innovative concept, the lead and final vehicle in the convoy alternate as fixed-reference differential GPS base stations. As the convoy moves out from a known location, the final vehicle remains behind to provide differential corrections to the GPS receivers in the rest of the vehicles. After traversing a predetermined distance in this fashion, the convoy is halted and the lead vehicle assumes the role of a differential reference station, providing enhanced accuracy to the trailing vehicle as it catches up to the pack. During this time, the lead vehicle takes advantage of on-site dwell to further improve the accuracy of its own fix. Once the last vehicle joins up with the rest, the base-station roles are reversed again, and the convoy resumes transit in "inchworm" fashion along its intended route. Disadvantages to this approach include the need for intermittent stops and the accumulating ambiguity in actual location of the appointed reference station.

Recall the Y-code chip rate is directly equal to the satellite cesium clock rate, or 10.23 MHz. Since the L1 carrier frequency of 1575.42 MHz is generated by multiplying the clock output by 154, there are consequently 154 carrier cycles for every Y-code chip. This implies even higher measurement precision is possible if the time of arrival is somehow referenced to the carrier instead of the pseudocode itself. Such codeless interferometric differential GPS schemes measure the phase of the L1 and L2 carrier frequencies to achieve centimeter accuracies, but they must start at a known geodetic location and typically require long dwell times. The Army's Engineer Topographic Laboratories (ETL) is in the process of developing a carrier-phase-differential system of this type that is expected to provide 1 to 3 centimeters (0.4 to 1.2 in) accuracy at a 60-Hz rate when finished sometime in 1996 [McPherson, 1991].

A reasonable extraction from the open literature of achievable position accuracies for the various GPS configurations is presented in Table 3.2. The Y code has dual-frequency estimation for atmospheric refraction and no S/A error component, so accuracies are better than stand-alone single-frequency C/A systems. Commercial DGPS accuracy, however, exceeds stand-alone military Y-code accuracy, particularly for small-area applications such as airports. Differential Y code is currently under consideration and may involve the use of a satellite to disseminate the corrections over a wide area.

A typical non-differential GPS was tested by Cooper and Durrant-White [1994] and yielded an accumulated position error of over 40 meters after extensive filtering.

Systems likely to provide the best accuracy are those that combine GPS with Inertial Navigation Systems (INS), because the INS position drift is bounded by GPS corrections [Motazed, 1993]. Similarly, the combination of GPS with odometry and a compass has been proposed by Byrne [1993].

In summary, the fundamental problems associated with using GPS for mobile robot navigation are as follows:
- Periodic signal blockage due to foliage and hilly terrain.
- Multi-path interference.
- Insufficient position accuracy for primary (stand-alone) navigation systems.

Arradondo-Perry [1992] provides a comprehensive listing of GPS receiver equipment, while Byrne [1993] presents a detailed evaluation of performance for five popular models.

Table 3.2: Summary of achievable position accuracies for various implementations of GPS.

GPS Implementation Method	Position Accuracy
C/A-code stand alone	100 m SEP (328 ft)
Y-code stand alone	16 m SEP (52 ft)
Differential (C/A-code)	3 m SEP (10 ft)
Differential (Y-code)	unknown (TBD)
Phase differential (codeless)	1 cm SEP (0.4 in)

3.2 Ground-Based RF Systems

Ground-based RF position location systems are typically of two types:

- Passive hyperbolic line-of-position phase-measurement systems that compare the time-of-arrival phase differences of incoming signals simultaneously emitted from surveyed transmitter sites.

- Active radar-like trilateration systems that measure the round-trip propagation delays for a number of fixed-reference transponders. Passive systems are generally preferable when a large number of vehicles must operate in the same local area, for obvious reasons.

3.2.1 Loran

An early example of the first category is seen in *Loran* (short for long range navigation). Developed at MIT during World War II, such systems compare the time of arrival of two identical signals broadcast simultaneously from high-power transmitters located at surveyed sites with a known separation baseline. For each finite time difference (as measured by the receiver) there is an associated hyperbolic line of position as shown in Figure 3.5. Two or more pairs of master/slave stations are required to get intersecting hyperbolic lines resulting in a two-dimensional (latitude and longitude) fix.

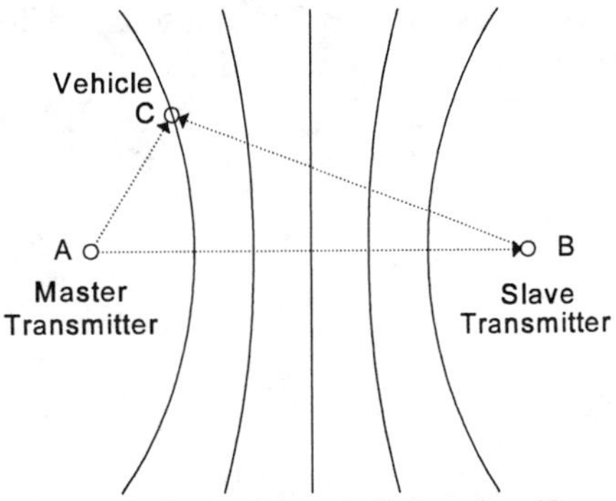

Figure 3.5: For each hyperbolic line-of-position, length ABC minus length AC equals some constant K. (Adapted from [Dodington, 1989].)

The original implementation (Loran A) was aimed at assisting convoys of liberty ships crossing the North Atlantic in stormy winter weather. Two 100 kW slave transmitters were located about 200 miles on either side of the master station. Non-line-of-sight ground-wave propagation at around 2 MHz was employed, with pulsed as opposed to continuous-wave transmissions to aid in sky-wave discrimination. The time-of-arrival difference was simply measured as the lateral separation of the two pulses on an oscilloscope display, with a typical accuracy of around 1 μs. This numerical value was matched to the appropriate line of position on a special Loran chart of the region, and the procedure then repeated for another set of transmitters. For discrimination purposes, four different frequencies were used, 50 kHz apart, with 24 different pulse repetition rates in the neighborhood of 20 to 35 pulses per second [Dodington, 1989]. In situations where the hyperbolic lines intersected more or less at right angles, the resulting (best-case) accuracy was about 1.5 kilometers.

Loran A was phased out in the early '80s in favor of Loran C, which achieves much longer over-the-horizon ranges through use of 5 MW pulses radiated from 400-meter (1300 ft) towers at a lower carrier frequency of 100 kHz. For improved accuracy, the phase differences of the first three cycles of the master and slave pulses are tracked by phase-lock-loops in the receiver and converted to a digital readout, which is again cross-referenced to a preprinted chart. Effective operational range is about 1000 miles, with best-case accuracies in the neighborhood of 100 meters

(330 ft). Coverage is provided by about 50 transmitter sites to all U.S. coastal waters and parts of the North Atlantic, North Pacific, and the Mediterranean.

3.2.2 Kaman Sciences *Radio Frequency Navigation Grid*

The Unmanned Vehicle Control Systems Group of Kaman Sciences Corporation, Colorado Springs, CO, has developed a scaled-down version of a Loran-type hyperbolic position-location system known as the *Radio Frequency Navigation Grid* (RFNG). The original application in the late 1970s involved autonomous route control of unmanned mobile targets used in live-fire testing of the laser-guided Copperhead artillery round [Stokes, 1989]. The various remote vehicles sense their position by measuring the phase differences in received signals from a master transmitter and two slaves situated at surveyed sites within a 30 km^2 (18.75 mi^2) area as shown in Figure 3.6. System resolution is 3 centimeters (1.5 in) at a 20 Hz update rate, resulting in a vehicle positioning repeatability of 1 meter (3.3 ft).

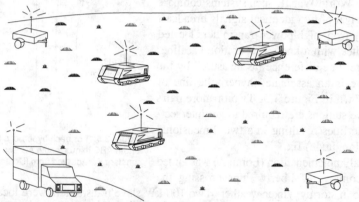

Figure 3.6: Kaman Sciences 1500 W navigation grid is a scaled-down version of the LORAN concept, covering an area 8 to 15 km on a side with a position-location repeatability of 1 m. (Courtesy of Kaman Sciences Corporation.)

Path trajectories are initially taught by driving a vehicle over the desired route and recording the actual phase differences observed. This file is then played back at run time and compared to measured phase difference values, with vehicle steering servoed in an appropriate manner to null any observed error signal. Velocity of advance is directly controlled by the speed of file playback. Vehicle speeds in excess of 50 km/h (30 mph) are supported over path lengths of up to 15 kilometers (9.4 mi) [Stokes, 1989]. Multiple canned paths can be stored and changed remotely, but vehicle travel must always begin from a known start point due to an inherent 6.3 meters (20 ft) phase ambiguity interval associated with the grid [Byrne et al., 1992].

The *Threat Array Control and Tracking Information Center* (TACTIC) is offered by Kaman Sciences to augment the RFNG by tracking and displaying the location and orientation of up to 24 remote vehicles [Kaman, 1991]. Real-time telemetry and recording of vehicle heading, position, velocity, status, and other designated parameters (i.e., fuel level, oil pressure, battery voltage) are supported at a 1 Hz update rate. The TACTIC operator has direct control over engine start, automatic path playback, vehicle pause/resume, and emergency halt functions. Non-line-of-sight operation is supported through use of a 23.825 MHz grid frequency in conjunction with a 72 MHz control and communications channel.

3.2.3 Precision Location Tracking and Telemetry System

Precision Technology, Inc., of Saline, MI, has recently introduced to the automotive racing world an interesting variation of the conventional phase-shift measurement approach (type 1 RF system). The company's *Precision Location* tracking and telemetry system employs a number of receive-only antennae situated at fixed locations around a racetrack to monitor a continuous sine wave transmission from a moving vehicle. By comparing the signals received by the various antennae to a common reference signal of identical frequency generated at the base station, relative changes in vehicle position with respect to each antenna can be inferred from resulting shifts in the respective phase relationships. The 58 MHz VHF signal allows for non-line-of-sight operation, with a resulting precision of approximately 1 to 10 centimeters (0.4 to 4 in) [Duchnowski, 1992]. From a robotics perspective, problems with this approach arise when more than one vehicle must be tracked. The system costs $200,000 to $400,000, depending on the number of receivers used. According to Duchnowski, the system is not suitable for indoor operations.

3.2.4 Motorola *Mini-Ranger Falcon*

An example of the active transponder category of ground-based RF position-location techniques is seen in the *Mini-Ranger Falcon* series of range positioning systems offered by the Government and Systems Technology Group of Motorola, Inc, Scottsdale, AZ [MOTOROLA]. The *Falcon 484* configuration depicted in Figure 3.7 is capable of measuring line-of-sight distances from 100 meters (328 ft) out to 75 kilometers (47 miles). An initial calibration is performed at a known location to determine the turn-around delay (TAD) for each transponder (i.e., the time required to transmit a response back to the interrogator after receipt of interrogation). The actual distance between the interrogator and a given transponder is found by [Byrne et al., 1992]:

$$D = \frac{(T_e - T_d)c}{2} \tag{3.1}$$

where
D = separation distance
T_e = total elapsed time
T_d = transponder turn-around delay
c = speed of light.

The MC6809-based range processor performs a least-squares position solution at a 1-Hz update rate, using range inputs from two, three, four, or 16 possible reference transponders.

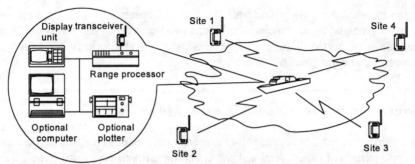

Figure 3.7: Motorola's *Mini-Ranger Falcon 484 R* position-location system provides 2 m (6.5 ft) accuracy over ranges of 100 m to 75 km (328 ft to 47 mi). (Courtesy of [MOTOROLA].)

The individual reference stations answer only to uniquely coded interrogations and operate in C-band (5410 to 5890 MHz) to avoid interference from popular X-band marine radars [Motorola, undated]. Up to 20 mobile users can time share the *Falcon 484* system (50 ms per user maximum). System resolution is in tenths of units (m, ft, or yd) with a range accuracy of 2 meters (6.5 ft) probable.

Power requirements for the fixed-location reference stations are 22 to 32 VDC at 13 W nominal, 8.5 W standby, while the mobile range processor and its associated transmitter-receiver and display unit draw 150 W at 22 to 32 VDC. The Falcon system comes in different, customized configurations. Complete system cost is $75,000 to $100,000.

3.2.5 Harris *Infogeometric* System

Harris Technologies, Inc., [HTI], Clifton, VA, is developing a ground-based R position location and communications strategy wherein moderately priced *infogeometric* (IG) devices cooperatively form self-organizing instrumentation and communication networks [Harris, 1994]. Each IG device in the network has full awareness of the identity, location, and orientation of all other IG devices and can communicate with other such devices in both party-line and point-to-point communication modes.

The IG devices employ digital code-division-multiple-access (CDMA) spread-spectrum R hardware that provides the following functional capabilities:
- Network level mutual autocalibration.
- Associative location and orientation tracking.
- Party-line and point-to-point data communications (with video and audio options).
- Distributed sensor data fusion.

Precision position location on the move is based on high-speed range trilateration from fixed reference devices, a method commonly employed in many instrumentation test ranges and other tracking system applications. In this approach, each beacon has an extremely accurate internal clock that is carefully synchronized with all other beacon clocks. A time-stamped (coded) R signal is periodically sent by each transmitter. The receiver is also equipped with a precision clock, so that it can compare the timing information and time of arrival of the incoming signals to its internal clock. This way, the system is able to accurately measure the signals' time of flight and thus the distance between the receiver and the three beacons. This method, known as "differential location regression" [Harris, 1994] is essentially the same as the locating method used in global positioning systems (GPS).

To improve accuracy over current range-lateration schemes, the HTI system incorporates mutual data communications, permitting each mobile user access to the time-tagged range measurements made by fixed reference devices and all other mobile users. This additional network-level range and timing information permits more accurate time synchronization among device clocks, and automatic detection and compensation for uncalibrated hardware delays.

Each omnidirectional CDMA spread-spectrum "geometric" transmission uniquely identifies the identity, location, and orientation of the transmitting source. Typically the available geometric measurement update rate is in excess of 1000 kHz. Harris quotes a detection radius of 500 meters (1640 ft) with 100 mW peak power transmitters. Larger ranges can be achieved with stronger transmitters. Harris also reports on "centimeter-class repeatability accuracy" obtained with a modified transmitter called an "Interactive Beacon." Tracking and communications at operating ranges of up to 20 kilometers (12.5 mi) are also supported by higher transmission power levels of 1 to 3 W. Typical "raw data" measurement resolution and accuracies are cited in Table 3.3.

Enhanced tracking accuracies for selected applications can be provided as cited in Table 3.4. This significant improvement in performance is provided by sensor data fusion algorithms that exploit the high degree of relational redundancy that is characteristic for infogeometric network measurements and communications.

Infogeometric enhancement algorithms also provide the following capabilities:

- Enhanced tracking in multipath and clutter — permits precision robotics tracking even when operating indoors.
- Enhanced near/far interference reduction — permits shared-spectrum operations in potentially large user networks (i.e., hundreds to thousands).

Table 3.3: Raw data measurement resolution and accuracy [Everett, 1995].

Parameter	Resolution	Biasing
Range	1	5 m
	3.3	16.4 ft
Bearing (Az, El)	2	2°
Orientation (Az)	2	2°

Table 3.4: Enhanced tracking resolution and accuracies obtained through sensor data fusion [Everett, 1995].

Parameter	Resolution	Biasing
Range	0.1 - 0.3	0.1 - 0.3m
	0.3 - 0.9	0.3 - 0.9ft
Bearing	0.5 - 1.0	0.5 - 1.0°
Orientation	0.5 - 1.0	0.5 - 1.0°

Operationally, mobile IG networks support precision tracking, communications, and command and control among a wide variety of potential user devices. A complete Infogeometric Positioning System is commercially available from [HTI], at a cost of $30,000 or more (depending on the number of transmitters required). In conversation with HTI we learned that the system requires an almost clear "line of sight" between the transmitters and receivers. In indoor applications, the existence of walls or columns obstructing the path will dramatically reduce the detection range and may result in erroneous measurements, due to multi-path reflections.

CHAPTER 4
SENSORS FOR MAP-BASED POSITIONING

Most sensors used for the purpose of map building involve some kind of distance measurement. There are three basically different approaches to measuring range:
- Sensors based on measuring the *time of flight* (TOF) of a pulse of emitted energy traveling to a reflecting object, then echoing back to a receiver.
- The *phase-shift measurement* (or *phase-detection*) ranging technique involves continuous wave transmission as opposed to the short pulsed outputs used in TOF systems.
- Sensors based on frequency-modulated (FM) radar. This technique is somewhat related to the (amplitude-modulated) phase-shift measurement technique.

4.1   Time-of-Flight Range Sensors

Many of today's range sensors use the *time-of-flight* (TOF) method. The measured pulses typically come from an ultrasonic, RF, or optical energy source. Therefore, the relevant parameters involved in range calculation are the speed of sound in air (roughly 0.3 m/ms — 1 ft/ms), and the speed of light (0.3 m/ns — 1 ft/ns). Using elementary physics, distance is determined by multiplying the velocity of the energy wave by the time required to travel the round-trip distance:

$$d = v t \qquad\qquad (4.1)$$

where
d = round-trip distance
v = speed of propagation
t = elapsed time.

The measured time is representative of traveling twice the separation distance (i.e., out and back) and must therefore be reduced by half to result in actual range to the target.

The advantages of TOF systems arise from the direct nature of their straight-line active sensing. The returned signal follows essentially the same path back to a receiver located coaxially with or in close proximity to the transmitter. In fact, it is possible in some cases for the transmitting and receiving transducers to be the same device. The absolute range to an observed point is directly available as output with no complicated analysis required, and the technique is not based on any assumptions concerning the planar properties or orientation of the target surface. The *missing parts* problem seen in triangulation does not arise because minimal or no offset distance between transducers is needed. Furthermore, TOF sensors maintain range accuracy in a linear fashion as long as reliable echo detection is sustained, while triangulation schemes suffer diminishing accuracy as distance to the target increases.

Potential error sources for TOF systems include the following:
- Variations in the speed of propagation, particularly in the case of acoustical systems.
- Uncertainties in determining the exact time of arrival of the reflected pulse.

- Inaccuracies in the timing circuitry used to measure the round-trip time of flight.
- Interaction of the incident wave with the target surface.

Each of these areas will be briefly addressed below, and discussed later in more detail.

a. Propagation Speed For mobile robotics applications, changes in the propagation speed of electromagnetic energy are for the most part inconsequential and can basically be ignored, with the exception of satellite-based position-location systems as presented in Chapter 3. This is not the case, however, for acoustically based systems, where the speed of sound is markedly influenced by temperature changes, and to a lesser extent by humidity. (The speed of sound is actually proportional to the square root of temperature in degrees Rankine.) An ambient temperature shift of just 30° F can cause a 0.3 meter (1 ft) error at a measured distance of 10 meters (35 ft) [Everett, 1985].

b. Detection Uncertainties So-called *time-walk errors* are caused by the wide dynamic range in returned signal strength due to varying reflectivities of target surfaces. These differences in returned signal intensity influence the rise time of the detected pulse, and in the case of fixed-threshold detection will cause the more reflective targets to appear closer. For this reason, *constant* fraction timing discriminators are typically employed to establish the detector threshold at some specified fraction of the peak value of the received pulse [Vuylsteke et al., 1990].

c. Timing Considerations Due to the relatively slow speed of sound in air, compared to light, acoustically based systems face milder timing demands than their light-based counterparts and as a result are less expensive. Conversely, the propagation speed of electromagnetic energy can place severe requirements on associated control and measurement circuitry in optical or RF implementations. As a result, TOF sensors based on the speed of light require sub-nanosecond timing circuitry to measure distances with a resolution of about a foot [Koenigsburg, 1982]. More specifically, a desired resolution of 1 millimeter requires a timing accuracy of 3 picoseconds (3×10^{-12} s) [Vuylsteke et al., 1990]. This capability is somewhat expensive to realize and may not be cost effective for certain applications, particularly at close range where high accuracies are required.

d. Surface Interaction When light, sound, or radio waves strike an object, any detected echo represents only a small portion of the original signal. The remaining energy reflects in scattered directions and can be absorbed by or pass through the target, depending on surface characteristics and the angle of incidence of the beam. Instances where no return signal is received at all can occur because of specular reflection at the object's surface, especially in the ultrasonic region of the energy spectrum. If the transmission source approach angle meets or exceeds a certain critical value, the reflected energy will be deflected outside of the sensing envelope of the receiver. In cluttered environments soundwaves can reflect from (multiple) objects and can then be received by other sensors. This phenomenon is known as crosstalk (see Figure 4.1). To compensate, repeated measurements are often averaged to bring the signal-to-noise ratio within acceptable levels, but at the expense of additional time required to determine a single range value. Borenstein and Koren [1995] proposed a method that allows individual sensors to detect and reject crosstalk.

Using this method much faster firing rates — under 100 ms for a complete scan with 12 sonars — are feasible.

4.1.1 Ultrasonic TOF Systems

Ultrasonic TOF ranging is today the most common technique employed on indoor mobile robotics systems, primarily due to the ready availability of low-cost systems and their ease of interface. Over the past decade, much research has been conducted investigating applicability in such areas as world modeling and collision avoidance, position estimation, and motion detection. Several researchers have more recently begun to assess the effectiveness of ultrasonic sensors in exterior settings [Pletta et

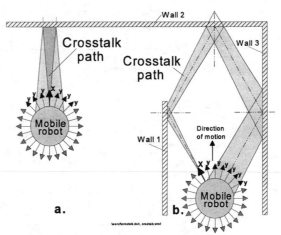

Figure 4.1: Crosstalk is a phenomenon in which one sonar picks up the echo from another. One can distinguish between a. direct crosstalk and b. indirect crosstalk.

al., 1992; Langer and Thorpe, 1992; Pin and Watanabe, 1993; Hammond, 1994]. In the automotive industry, BMW now incorporates four piezoceramic transducers (sealed in a membrane for environmental protection) on both front and rear bumpers in its Park Distance Control system [Siuru, 1994]. A detailed discussion of ultrasonic sensors and their characteristics with regard to indoor mobile robot applications is given in [Jörg, 1994].

Two of the most popular commercially available ultrasonic ranging systems will be reviewed in the following sections.

4.1.1.1 Massa Products Ultrasonic Ranging Module Subsystems

Massa Products Corporation, Hingham, MA, offers a full line of ultrasonic ranging subsystems with maximum detection ranges from 0.6 to 9.1 meters (2 to 30 ft) [MASSA]. The *E-201B series* sonar operates in the bistatic mode with separate transmit and receive transducers, either side by side for echo ranging or as an opposed pair for unambiguous distance measurement between two uniquely defined points. This latter configuration is sometimes used in ultrasonic position location systems and provides twice the effective operating range with respect to that advertised for conventional echo ranging. The *E-220B series* (see Figure 4.2) is designed for monostatic (single transducer) operation but is otherwise functionally identical to the *E-201B*. Either version can be externally triggered on command, or internally triggered by a free-running oscillator at a repetition rate determined by an external resistor (see Figure 4.3).

Selected specifications for the four operating frequencies available in the *E-220B series* are listed in Table 4.1 below. A removable focusing horn is provided for the 26- and 40-kHz models that decreases the effective beamwidth (when installed) from 35 to 15 degrees. The horn must be in place to achieve the maximum listed range.

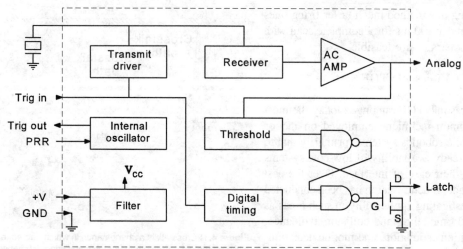

Figure 4.2: The single-transducer Massa *E-220B-series* ultrasonic ranging module can be internally or externally triggered, and offers both analog and digital outputs. (Courtesy of Massa Products Corp.)

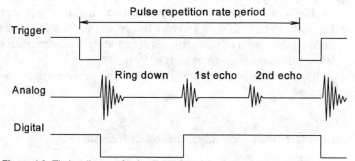

Figure 4.3: Timing diagram for the *E-220B series* ranging module showing analog and digital output signals in relationship to the trigger input. (Courtesy of Massa Products Corp.)

Table 4.1: Specifications for the monostatic E-220B Ultrasonic Ranging Module Subsystems. The E-201 series is a bistatic configuration with very similar specifications. (Courtesy of Massa Products Corp.)

Parameter	E-220B/215	E-220B/150	E-220B/40	E-220B/26	Units
Range	10 - 61	20 - 152	61 - 610	61 - 914	cm
	4 - 24	8 - 60	24 - 240	24 - 360	in
Beamwidth	10	10	35 (15)	35 (15)	°
Frequency	215	150	40	26	kHz
Max rep rate	150	100	25	20	Hz
Resolution	0.076	0.1	0.76	1	cm
	0.03	0.04	0.3	0.4	in
Power	8 - 15	8 - 15	8 - 15	8 - 15	VDC
Weight	4 - 8	4 - 8	4 - 8	4 - 8	oz

4.1.1.2 Polaroid Ultrasonic Ranging Modules

The Polaroid ranging module is an active TOF device developed for automatic camera focusing, which determines the range to target by measuring elapsed time between the transmission of an ultrasonic waveform and the detected echo [Biber et al., 1987, POLAROID]. This system is the most widely found in mobile robotics literature [Koenigsburg, 1982; Moravec and Elfes, 1985; Everett, 1985; Kim, 1986; Moravec, 1988; Elfes, 1989; Arkin, 1989; Borenstein and Koren, 1990; 1991; 1994;

Figure 4.4: The Polaroid OEM kit included the transducer and a small electronics interface board.

Borenstein et al., 1995], and is representative of the general characteristics of such ranging devices. The most basic configuration consists of two fundamental components: 1) the ultrasonic transducer, and 2) the ranging module electronics. Polaroid offers OEM kits with two transducers and two ranging module circuit boards for less than $100 (see Figure 4.4).

A choice of transducer types is now available. In the original instrument-grade electrostatic version, a very thin metal diaphragm mounted on a machined backplate formed a capacitive transducer as illustrated in Figure 4.5 [POLAROID, 1991]. The system operates in the monostatic transceiver mode so that only a single transducer is necessary to acquire range data. A smaller diameter electrostatic transducer (*7000-series*) has also been made available, developed for the Polaroid *Spectra* camera [POLAROID, 1987]. A more rugged piezoelectric (*9000-series*) *environmental transducer* for applications in severe environmental conditions including vibration is able to meet or exceed the SAE J1455 January 1988 specification for heavy-duty trucks. Table 4.2 lists the technical specifications for the different Polaroid transducers.

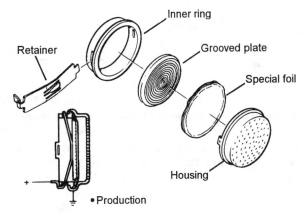

Figure 4.5: The Polaroid instrument grade electrostatic transducer consists of a gold-plated plastic foil stretched across a machined backplate. (Reproduced with permission from Polaroid [1991].)

The original Polaroid ranging module functioned by transmitting a *chirp* of four discrete frequencies at about of 50 kHz. The *SN28827* module was later developed with reduced parts count, lower power consumption, and simplified computer interface requirements. This second-generation board transmits only a single frequency at 49.1 kHz. A third-generation board (*6500 series*) introduced in 1990 provided yet a further reduction in interface circuitry, with the ability to detect and report multiple echoes [Polaroid, 1990]. An *Ultrasonic Ranging Developer's Kit* based on the Intel *80C196* microprocessor is now available for use with the *6500 series* ranging module that allows software control of transmit frequency, pulse width, blanking time, amplifier gain, and maximum range [Polaroid, 1993].

The range of the Polaroid system runs from about 41 centimeters to 10.5 meters (1.33 ft to 35 ft). However, using custom circuitry suggested in [POLAROID, 1991] the minimum range can be reduced reliably to about 20 centimeters (8 in) [Borenstein et al., 1995]. The beam dispersion angle is approximately 30 degrees. A typical operating cycle is as follows.

1. The control circuitry fires the transducer and waits for indication that transmission has begun.
2. The receiver is blanked for a short period of time to prevent false detection due to ringing from residual transmit signals in the transducer.
3. The received signals are amplified with increased gain over time to compensate for the decrease in sound intensity with distance.
4. Returning echoes that exceed a fixed threshold value are recorded and the associated distances calculated from elapsed time.

Table 4.2: Specifications for the various Polaroid ultrasonic ranging modules. (Courtesy of Polaroid.)

Parameter	Original	SN28827	6500	Units
Maximum range	10.5	10.5	10.5	m
	35	35	35	ft
Minimum range*	25	20	20	cm
	10.5	6	6	in
Number of pulses	56	16	16	
Blanking time	1.6	2.38	2.38	ms
Resolution	1	2	1	%
Gain steps	16	12	12	
Multiple echo	no	yes	yes	
Programmable frequency	no	no	yes	
Power	4.7 - 6.8	4.7 - 6.8	4.7 - 6.8	V
	200	100	100	mA

* with custom electronics (see [Borenstein et al., 1995].)

Figure 4.6 [Polaroid, 1990] illustrates the operation of the sensor in a timing diagram. In the *single-echo* mode of operation for the *6500-series* module, the *blank* (BLNK) and *blank-inhibit* (BINH) lines are held low as the *initiate* (INIT) line goes high to trigger the outgoing pulse train. The *internal blanking* (BLANKING) signal automatically goes high for 2.38 milliseconds to prevent transducer ringing from being misinterpreted as a returned echo. Once a valid return is received, the echo (ECHO) output will latch high until reset by a high-to-low transition on INIT.

For multiple-echo processing, the *blanking* (BLNK) input must be toggled high for at least 0.44 milliseconds after detection of the first return signal to reset the *echo* output for the next return.

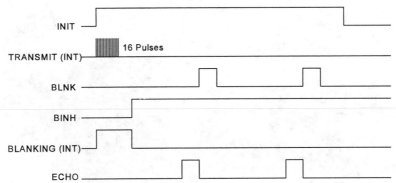

Figure 4.6: Timing diagram for the 6500-*Series Sonar Ranging Module* executing a multiple-echo-mode cycle with blanking input. (Courtesy of Polaroid Corp.)

4.1.2 Laser-Based TOF Systems

Laser-based TOF ranging systems, also known as *laser radar* or *lidar*, first appeared in work performed at the Jet Propulsion Laboratory, Pasadena, CA, in the 1970s [Lewis and Johnson, 1977]. Laser energy is emitted in a rapid sequence of short bursts aimed directly at the object being ranged. The time required for a given pulse to reflect off the object and return is measured and used to calculate distance to the target based on the speed of light. Accuracies for early sensors of this type could approach a few centimeters over the range of 1 to 5 meters (3.3 to 16.4 ft) [NASA, 1977; Depkovich and Wolfe, 1984].

4.1.2.1 Schwartz Electro-Optics Laser Rangefinders

Schwartz Electro-Optics, Inc. (SEO), Orlando, FL, produces a number of laser TOF rangefinding systems employing an innovative time-to-amplitude-conversion scheme to overcome the sub-nanosecond timing requirements necessitated by the speed of light. As the laser fires, a precision capacitor begins discharging from a known set point at a constant rate. An analog-to-digital conversion is performed on the sampled capacitor voltage at the precise instant a return signal is detected, whereupon the resulting digital representation is converted to range using a look-up table.

SEO *LRF-200 OEM Laser Rangefinders*
The *LRF-200 OEM Laser Rangefinder* shown in Figure 4.7 features compact size, high-speed processing, and the ability to acquire range information from most surfaces (i.e., minimum 10-percent Lambertian reflectivity) out to a maximum of 100 meters (328 ft). The basic system uses a pulsed InGaAs laser diode in conjunction with an avalanche photodiode detector, and is available

with both analog and digital (RS-232) outputs. Table 4.3 lists general specifications for the sensor's performance [SEO, 1995a].

Figure 4.7: The *LRF-200 OEM Laser Rangefinder*. (Courtesy of Schwartz Electro-Optics, Inc.)

Another adaptation of the LRF-200 involved the addition of a mechanical single-DOF beam scanning capability. Originally developed for use in submunition sensor research, the *Scanning Laser Rangefinder* is currently installed on board a remotely piloted vehicle. For this application, the sensor is positioned so the forward motion of the RPV is perpendicular to the vertical scan plane, since three-dimensional target profiles are required [SEO, 1991b]. In a second application, the *Scanning Laser Rangefinder* was used by the Field Robotics Center at Carnegie Mellon University as a terrain mapping sensor on their unmanned autonomous vehicles.

Table 4.3: Selected specifications for the *LRF 200 OEM Laser Rangefinder*. (Courtesy of Schwartz Electro-Optics, Inc.)

Parameter	Value	Units
Range (non-cooperative target)	1 to 100	m
	3.3-328	ft
Accuracy	±30	cm
	±12	in
Range jitter	±12	cm
	±4.7	in
Wavelength	902	nm
Diameter	89	mm
	3.5	in
Length	178	mm
	7	in
Weight	1	kg
	2.2	lb
Power	8 to 24	VDC
	5	W

Table 4.4: Selected specifications for the SEO *Scanning Laser Rangefinder*. (Courtesy of Schwartz Electro-Optics, Inc.)

Parameter	Value	Units
Range	1-100	m
	3.3-330	ft
Accuracy	±30	cm
	±12	in
Scan angle	±30	°
Scan rate	24.5- 30.3	kHz
Samples per scan	175	
Wavelength	920	nm
Diameter	127	mm
	5	in
Length	444	mm
	17.5	in
Weight	5.4	kg
	11.8	lb
Power	8-25	VDC

SEO *Scanning Helicopter Interference Envelope Laser Detector (SHIELD)*

This system was developed for the U.S. Army [SEO, 1995b] as an onboard pilot alert to the presence of surrounding obstructions in a 60-meter radius hemispherical envelope below the helicopter. A high-pulse-repetition-rate GaAs eye-safe diode emitter shares a common aperture with a sensitive avalanche photodiode detector. The transmit and return beams are reflected from a motor-driven prism rotating at 18 rps (see Figure 4.9). Range measurements are correlated with the azimuth angle using an optical encoder. Detected obstacles are displayed on a 5.5-inch color monitor. Table 4.5 lists the key specifications of the *SHIELD*.

Table 4.5: Selected specifications for the *Scanning Helicopter Interference Envelope Laser Detector (SHIELD)*. (Courtesy of Schwartz Electro-Optics, Inc.)

Parameter	Value	Units
Maximum range	>60	m
(hemispherical envelope)	>200	ft
Accuracy	<30	cm
	1	ft
Wavelength	905	nm
Scan angle	360	°
Scan rate	18	Hz
Length	300	mm
	11.75	in
Weight	15	lb
Power	18	VDC
	<5	A

Figure 4.8: The *Scanning Helicopter Interference Envelope Laser Detector (SHIELD)*. (Courtesy of Schwartz Electro-Optics, Inc.)

SEO *TreeSense*

The *TreeSense* system was developed by SEO for automating the selective application of pesticides to orange trees, where the goal was to enable individual spray nozzles only when a tree was detected within their associated field of coverage. The sensing subsystem (see Figure 4.9) consists of a horizontally oriented unit mounted on the back of an agricultural vehicle, suitably equipped with a rotating mirror arrangement that scans the beam in a vertical plane orthogonal to the direction of travel. The scan rate is controllable up to 40 rps (35 rps typical). The ranging subsystem is gated on and off twice during each revolution to illuminate two 90-degree fan-shaped sectors to a maximum range of 7.6 meters (25 ft) either side of the vehicle as shown in Figure 4.10. The existing hardware is theoretically capable of ranging to 9 meters (30 ft) using a PIN photodiode and can be extended further through an upgrade option that incorporates an avalanche photodiode detector.

The *TreeSense* system is hard-wired to a valve manifold to enable/disable a vertical array of nozzles for the spraying of insecticides, but analog as well as digital (RS-232) output can easily be made available for other applications. The system is housed in a rugged aluminum enclosure with a total weight of only 2.2 kilograms (5 lb). Power requirements are 12 W at 12 VDC. Further details on the system are contained in Table 4.6.

Figure 4.9: The SEO *TreeSense*. (Courtesy of Schwartz Electro-Optics, Inc.)

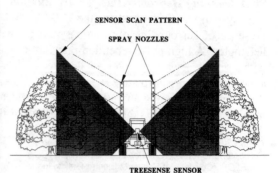

Figure 4.10: Scanning pattern of the SEO *TreeSense* system. (Courtesy of Schwartz Electro-Optics, Inc.)

SEO *AutoSense*

The *AutoSense I* system was developed by SEO under a Department of Transportation Small Business Innovative Research (SBIR) effort as a replacement for buried inductive loops for traffic signal control. (Inductive loops don't always sense motorcyclists and some of the smaller cars with fiberglass or plastic body panels, and replacement or maintenance can be expensive as well as disruptive to traffic flow.) The system is configured to look down at about a 30-degree angle on moving vehicles in a traffic lane as illustrated in Figure 4.12.

AutoSense I uses a PIN photo-diode detector and a pulsed (8 ns) InGaAs near-infrared laser-diode source with peak power of 50 W. The laser output is directed by a beam splitter into a pair of cylindrical lenses to generate two fan-shaped beams 10 degrees apart in elevation for improved target detection. (The original prototype projected only a single spot of light, but ran into problems due to target absorption and specular reflection.) As an added benefit, the use of two separate beams makes it possible to calculate the speed of moving vehicles to an accuracy of 1.6 km/h (1 mph). In addition, a two-dimensional image (i.e., length and width) is formed of each vehicle as it passes through the sensor's field of view,

Table 4.6: Selected specifications for the *TreeSense* system. (Courtesy of Schwartz Electro-Optics, Inc.)

Parameter	Value	Units
Maximum range	9	m
	30	ft
Accuracy (in % of measured range)	1	%
Wavelength	902	nm
Pulse repetition frequency	15	KHz
Scan rate	29.3	rps
Length	229	mm
	9	in
Width	229	mm
	9	in
Height	115	mm
	4.5	in
Weight	5	lbs
Power	12	V
	12	W

Figure 4.11: Color-coded range image created by the SEO *TreeSense* system. (Courtesy of Schwartz Electro-Optics, Inc.)

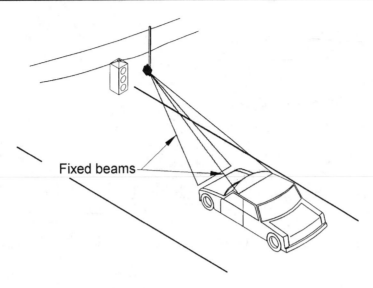

Figure 4.12: Two fan-shaped beams look down on moving vehicles for improved target detection. (Courtesy of Schwartz Electro-Optics, Inc.)

opening the door for numerous vehicle classification applications under the Intelligent Vehicle Highway Systems concept.

AutoSense II is an improved second-generation unit (see Figure 4.13) that uses an avalanche photodiode detector instead of the PIN photodiode for greater sensitivity, and a multi-faceted rotating mirror with alternating pitches on adjacent facets to create the two beams. Each beam is scanned across the traffic lane 720 times per second, with 15 range measurements made per scan. This azimuthal scanning action generates a precise three-dimensional profile to better facilitate vehicle classification in automated toll booth applications. An abbreviated system block diagram is depicted in Figure 4.14.

Figure 4.13: *The AutoSense II* is SEO's active-infrared overhead vehicle imaging sensor. (Courtesy of Schwartz Electro-Optics, Inc.)

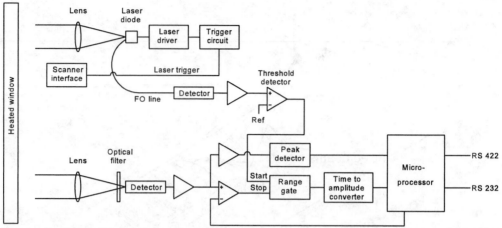

Figure 4.14: Simplified block diagram of the *AutoSense II* time-of-flight 3-D ranging system. (Courtesy of Schwartz Electro-Optics, Inc.)

Intensity information from the reflected signal is used to correct the "time-walk" error in threshold detection resulting from varying target reflectivities, for an improved range accuracy of 7.6 cm (3 in) over a 1.5 to 15 m (5 to 50 ft) field of regard. The scan resolution is 1 degree, and vehicle velocity can be calculated with an accuracy of 3.2 km/h (2 mph) at speeds up to 96 km/h (60 mph). A typical scan image created with the Autosense II is shown in Figure 4.15.

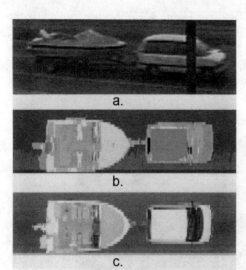

A third-generation *AutoSense III* is now under development for an application in Canada that requires 3-dimensional vehicle profile generation at speeds up to 160 km/h (100 mph). Selected specifications for the AutoSense II package are provided in Table 4.7.

Figure 4.15: Output sample from a scan with the *AutoSense II*.
a. Actual vehicle with trailer (photographed with a conventional camera).
b. Color-coded range information.
c. Intensity image.
(Courtesy of Schwartz Electro-Optics, Inc.)

Table 4.7: Selected specifications for the AutoSense II ranging system. (Courtesy of Schwartz Electro-Optics, Inc.)

Parameter	Value	Units
Range	0.61-1.50	m
	2-50	ft
Accuracy	7.5	cm
	3	in
Wavelength	904	nm
Pulse repetition rate	86.4	kHz
Scan rate	720	scans/s/scanline
Range measurements per scan	30	
Weight	11.4	kg
	25	lb
Power	115	VAC
	75	W

4.1.2.2 RIEGL Laser Measurement Systems

RIEGL Laser Measurement Systems, Horn, Austria, offers a number of commercial products (i.e., laser binoculars, surveying systems, "speed guns," level sensors, profile measurement systems, and tracking laser scanners) employing short-pulse TOF laser ranging. Typical applications include lidar altimeters, vehicle speed measurement for law enforcement, collision avoidance for cranes and vehicles, and level sensing in silos. All RIEGL products are distributed in the United States by RIEGEL USA, Orlando, FL.

LD90-3 Laser Rangefinder
The RIEGL *LD90-3 series* laser rangefinder (see Figure 4.16) employs a near-infrared laser diode source and a photodiode detector to perform TOF ranging out to 500 meters (1,640 ft) with diffuse surfaces, and to over 1,000 meters (3,281 ft) in the case of co-operative targets. Round-trip propagation time is precisely measured by a quartz-stabilized clock and converted to measured distance by an internal microprocessor using one of two available algorithms. The clutter suppression algorithm incorporates a combination of range measurement averaging and noise rejection techniques to filter out backscatter from airborne particles, and is therefore useful when operating under conditions of poor visibility [Riegl, 1994]. The *standard measurement* algorithm, on the other hand, provides rapid range measurements without regard for noise suppression, and can subsequently deliver a higher update rate under more favorable environmental conditions. Worst-case range measurement accuracy is ±5 centimeters (±2 in), with typical values of around ±2 centimeters (±0.8 in). See Table 4.8 for a complete listing of the LD90-3's features.

The pulsed near-infrared laser is Class-1 eye safe under all operating conditions. A nominal beam divergence of 0.1 degrees (2 mrad) for the LD90-3100 unit (see Tab. 4.9 below) produces a 20 centimeter (8 in) footprint of illumination at 100 meters (328 ft) [Riegl, 1994]. The complete system is housed in a small light-weight metal enclosure weighing only 1.5 kilograms (3.3 lb), and

Figure 4.16: The RIEGL *LD90-3 series* laser rangefinder. (Courtesy of Riegl USA.)

draws 10 W at 11 to 18 VDC. The standard output format is serial RS-232 at programmable data rates up to 19.2 kilobits per second, but RS-422 as well as analog options (0 to 10 VDC and 4 to 20 mA current-loop) are available upon request.

Table 4.8: Selected specifications for the RIEGL LD90-3 series laser rangefinder. (Courtesy of RIEGL Laser Measurement Systems.)

Parameter		LD90-3100	LD90-3300	Units
Maximum range	(diffuse)	150	400	m
		492	1,312	ft
	(cooperative)	>1000	>1000	m
		>3,280	>3,280	ft
Minimum range		1	3-5	m
Accuracy	(distance)	2	5	cm
		¾	2	in
	(velocity)	0.3	0.5	m/s
Beam divergence		2	2.8	mrad
Output	(digital)	RS-232, -422	RS-232, -422	
	(analog)	0-10	0-10	VDC
Power		11-18	11-18	VDC
		10	10	W
Size		22×13×7.6	22×13×7.6	cm
		8.7×5.1×3	8.7×5.1×3	in
Weight		3.3	3.3	lb

Scanning Laser Rangefinders

The *LRS90-3 Laser Radar Scanner* is an adaptation of the basic LD90-3 electronics, fiber-optically coupled to a remote scanner unit as shown in Figure 4.17. The scanner package contains no internal electronics and is thus very robust under demanding operating conditions typical of industrial or robotics scenarios. The motorized scanning head pans the beam back and forth in the horizontal plane at a 10-Hz rate, resulting in 20 data-gathering sweeps per second. Beam divergence is 0.3 degrees (5 mrad) with the option of expanding in the vertical direction if desired up to 2 degrees.

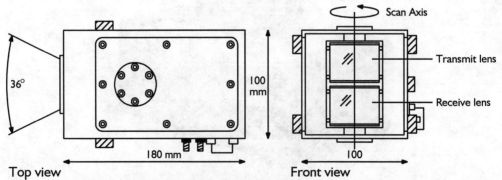

Figure 4.17: The *LRS90-3 Laser Radar Scanner* consists of an electronics unit (not shown) connected via a duplex fiber-optic cable to the remote scanner unit depicted above. (Courtesy of RIEGL USA.)

The *LSS390 Laser Scanning System* is very similar to the LRS90-3, but scans a more narrow field of view (10°) with a faster update rate (2000 Hz) and a more tightly focused beam. Range accuracy is 10 centimeters (4 in) typically and 20 centimeters (8 in) worst case. The LSS390 unit is available with an RS-422 digital output (19.2 kbs standard, 150 kbs optional) or a 20 bit parallel TTL interface.

Table 4.9: Typical specifications for the *LRS90-3 Laser Radar Scanner* and the *LSS390 Laser Scanner System.* (Courtesy of RIEGL USA.)

Parameter		LRS90-3	LSS390	Units
Maximum range		80	60	m
		262	197	ft
Minimum range		2	1	m
		6.5	3.25	ft
Accuracy		3	10	cm
		1.2	4	ft
Beam divergence		5	3.5	mrad
Sample rate		1000	2000	Hz
Scan range		18	10	°
Scan rate		10	10	scans/s
Output	(digital)	RS-232, -422	parallel, RS-422	
Power		11-15	9-16	VDC
		880		mA
Size	(electronics)	22×13×7.6	22×13×7.6	cm
		8.7×5.1×3	8.7×5.1×3	in
	(scanner)	18×10×10	18×10×10	cm
		7×4×4	7×4×4	in
Weight	(electronics)	7.25	2.86	lb
	(scanner)	3.52	2	lb

4.1.2.3 RVSI Long Optical Ranging and Detection System

Robotic Vision Systems, Inc., Haupaugue, NY, has conceptually designed a laser-based TOF ranging system capable of acquiring three-dimensional image data for an entire scene without scanning. The *Long Optical Ranging and Detection System (LORDS)* is a patented concept incorporating an optical encoding technique with ordinary vidicon or solid state camera(s), resulting in precise distance measurement to multiple targets in a scene illuminated by a single laser pulse. The design configuration is relatively simple and comparable in size and weight to traditional TOF and phase-shift measurement laser rangefinders (Figure 4.18).

Major components will include a single laser-energy source; one or more imaging cameras, each with an electronically implemented shuttering mechanism; and the associated control and processing electronics. In a typical configuration, the laser will emit a 25-mJ (millijoule) pulse lasting 1 nanosecond, for an effective transmission of 25 mW. The anticipated operational wavelength will lie between 532 and 830 nanometers, due to the ready availability within this range of the required laser source and imaging arrays.

The cameras will be two-dimensional CCD arrays spaced closely together with parallel optical axes resulting in nearly identical, multiple views of the illuminated surface. Lenses for these

Figure 4.18: Simplified block diagram of a three-camera configuration of the *LORDS* 3-D laser TOF rangefinding system. (Courtesy of Robotics Vision Systems, Inc.)

cameras will be of the standard photographic varieties between 12 and 135 millimeters. The shuttering function will be performed by microchannel plate image intensifiers (MCPs) 18 or 25 millimeters in size, which will be gated in a binary encoding sequence, effectively turning the CCDs on and off during the detection phase. Control of the system will be handled by a single-board processor based on the Motorola *MC-68040*.

LORDS obtains three-dimensional image information in real time by employing a novel time-of-flight technique requiring only a single laser pulse to collect all the information for an entire scene. The emitted pulse journeys a finite distance over time; hence, light traveling for 2 milliseconds will illuminate a scene further away than light traveling only 1 millisecond.

The entire sensing range is divided into discrete distance increments, each representing a distinct range plane. This is accomplished by simultaneously gating the MCPs of the observation cameras according to their own unique on-off encoding pattern over the duration of the detection phase. This binary gating alternately blocks and passes any returning reflection of the laser emission off objects within the field-of-view. When the gating cycles of each camera are lined up and compared, there exists a uniquely coded correspondence which can be used to calculate the range to any pixel in the scene.

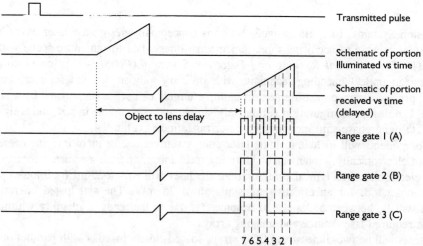

Figure 4.19: Range ambiguity is reduced by increasing the number of binary range gates. (Courtesy of Robotic Vision Systems, Inc.)

For instance, in a system configured with only one camera, the gating MCP would be cycled on for half the detection duration, then off the remainder of the time. Figure 4.19 shows any object detected by this camera must be positioned within the first half of the sensor's overall range (half the distance the laser light could travel in the allotted detection time). However, significant distance ambiguity exists because the exact time of detection of the reflected energy could have occurred anywhere within this relatively long interval.

This ambiguity can be reduced by a factor of two through the use of a second camera with its associated gating cycled at twice the rate of the first. This scheme would create two complete *on-off* sequences, one taking place while the first camera is on and the other while the first camera is off. Simple binary logic can be used to combine the camera outputs and further resolve the range. If the first camera did not detect an object but the second did, then by examining the instance when the first camera is off and the second is on, the range to the object can be associated with a relatively specific time frame. Incorporating a third camera at again twice the gating frequency (i.e., two cycles for every one of camera two, and four cycles for every one of camera one) provides even more resolution. As Figure 4.20 shows, for each additional CCD array incorporated into the system, the number of distance divisions is effectively doubled.

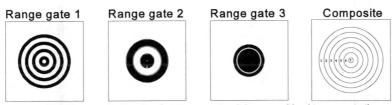

Figure 4.20: Binary coded images from range gates 1-3 are combined to generate the composite range map on the far right. (Courtesy of Robotics Vision Systems, Inc.)

Alternatively, the same encoding effect can be achieved using a single camera when little or no relative motion exists between the sensor and the target area. In this scenario, the laser is pulsed multiple times, and the gating frequency for the single camera is sequentially changed at each new transmission. This creates the same detection intervals as before, but with an increase in the time required for data acquisition.

LORDS is designed to operate over distances between one meter and several kilometers. An important characteristic is the projected ability to range over selective segments of an observed scene to improve resolution in that the depth of field over which a given number of range increments is spread can be variable. The entire range of interest is initially observed, resulting in the maximum distance between increments (coarse resolution). An object detected at this stage is thus localized to a specific, abbreviated region of the total distance.

The sensor is then electronically reconfigured to cycle only over this region, which significantly shortens the distance between increments, thereby increasing resolution. A known delay is introduced between transmission and the time when the detection/gating process is initiated. The laser light thus travels to the region of interest without concern for objects positioned in the foreground.

4.2 Phase-Shift Measurement

The *phase-shift measurement* (or *phase-detection*) ranging technique involves continuous wave transmission as opposed to the short pulsed outputs used in TOF systems. A beam of amplitude-modulated laser, RF, or acoustical energy is directed towards the target. A small portion of this wave (potentially up to six orders of magnitude less in amplitude) is reflected by the object's surface back to the detector along a direct path [Chen et al., 1993]. The returned energy is compared to a simultaneously generated reference that has been split off from the original signal, and the relative phase shift between the two is measured as illustrated in Figure 4.21 to ascertain the round-trip distance the wave has traveled. For high-frequency RF- or laser-based systems, detection is usually preceded by heterodyning the reference and received signals with an intermediate frequency (while preserving the relative phase shift) to allow the phase detector to operate at a more convenient lower frequency [Vuylsteke, 1990].

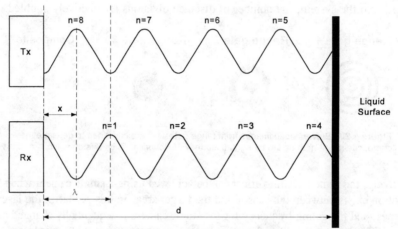

Figure 4.21: Relationship between outgoing and reflected waveforms, where x is the distance corresponding to the differential phase. (Adapted from [Woodbury et al., 1993].)

The relative phase shift expressed as a function of distance to the reflecting target surface is [Woodbury et al., 1993]:

$$\phi = \frac{4\pi d}{\lambda} \tag{4.1}$$

where
ϕ = phase shift
d = distance to target
λ = modulation wavelength.

The desired distance to target d as a function of the measured phase shift ϕ is therefore given by

$$d = \frac{\phi\lambda}{4\pi} = \frac{\phi c}{4\pi f} \tag{4.2}$$

where
f = modulation frequency.

    For square-wave modulation at the relatively low frequencies typical of ultrasonic systems (20 to 200 kHz), the phase difference between incoming and outgoing waveforms can be measured with the simple linear circuit shown in Figure 4.22 [Figueroa and Barbieri, 1991]. The output of the *exclusive-or* gate goes high whenever its inputs are at opposite logic levels, generating a voltage across capacitor C that is proportional to the phase shift. For example, when the two signals are in phase (i.e., $\phi = 0$), the gate output stays low and V is zero; maximum output voltage occurs when ϕ reaches 180 degrees. While easy to implement, this simplistic approach is limited to very low frequencies, and may require frequent calibration to compensate for drifts and offsets due to component aging or changes in ambient conditions [Figueroa and Lamancusa, 1992].

    At higher frequencies, the phase shift between outgoing and reflected sine waves can be measured by multiplying the two signals together in an electronic mixer, then averaging the product over many modulation cycles [Woodbury et al., 1993]. This integration process can be relatively time consuming, making it difficult to achieve extremely rapid update rates. The result can be expressed mathematically as follows [Woodbury et al., 1993]:

$$\lim_{T \to \infty} \frac{1}{T} \int_0^T \sin\left(\frac{2\pi c}{\lambda}t + \frac{4\pi d}{\lambda}\right) \sin\left(\frac{2\pi c}{\lambda}\right) dt \tag{4.3}$$

which reduces to

$$A\cos\frac{4\pi d}{\lambda} \tag{4.4}$$

where
t = time
T = averaging interval
A = amplitude factor from gain of integrating amplifier.

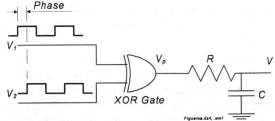

Figure 4.22: At low frequencies typical of ultrasonic systems, a simple phase-detection circuit based on an *exclusive-or* gate will generate an analog output voltage proportional to the phase difference seen by the inputs. (Adapted from [Figueroa and Barbieri, 1991].)

    From the earlier expression for ϕ, it can be seen that the quantity actually measured is in fact the *cosine* of the phase shift and not the phase shift itself [Woodbury et al., 1993]. This situation introduces a so-called *ambiguity interval* for scenarios where the round-trip distance exceeds the modulation wavelength (i.e., the phase measurement becomes ambiguous once ϕ exceeds 360°). Conrad and Sampson [1990] define this ambiguity interval as the maximum range that allows the phase difference to go through one complete cycle of 360 degrees:

$$R_a = \frac{c}{2f}$$
(4.5)

where
R_a = ambiguity range interval
f = modulation frequency
c = speed of light.

Referring again to Figure 4.21, it can be seen that the total round-trip distance $2d$ is equal to some integer number of wavelengths $n\lambda$ plus the fractional wavelength distance x associated with the phase shift. Since the cosine relationship is not single valued for all of ϕ, there will be more than one distance d corresponding to any given phase shift measurement [Woodbury et al., 1993]:

$$\cos\phi = \cos\left(\frac{4\pi d}{\lambda}\right) = \cos\left(\frac{2\pi(x + n\lambda)}{\lambda}\right)$$
(4.6)

where:
$d = (x + n\lambda) / 2$ = true distance to target.
x = distance corresponding to differential phase ϕ.
n = number of complete modulation cycles.

The potential for erroneous information as a result of this *ambiguity interval* reduces the appeal of phase-detection schemes. Some applications simply avoid such problems by arranging the optical path so that the maximum possible range is within the ambiguity interval. Alternatively, successive measurements of the same target using two different modulation frequencies can be performed, resulting in two equations with two unknowns, allowing both x and n to be uniquely determined. Kerr [1988] describes such an implementation using modulation frequencies of 6 and 32 MHz.

Advantages of continuous-wave systems over pulsed time-of-flight methods include the ability to measure the direction and velocity of a moving target in addition to its range. In 1842, an Austrian by the name of Johann Doppler published a paper describing what has since become known as the *Doppler effect*. This well-known mathematical relationship states that the frequency of an energy wave reflected from an object in motion is a function of the relative velocity between the object and the observer. This subject was discussed in detail in Chapter 1.

As with TOF rangefinders, the paths of the source and the reflected beam are coaxial for phase-shift-measurement systems. This characteristic ensures objects cannot cast shadows when illuminated by the energy source, preventing the *missing parts* problem. Even greater measurement accuracy and overall range can be achieved when cooperative targets are attached to the objects of interest to increase the power density of the return signal.

Laser-based continuous-wave (CW) ranging originated out of work performed at the Stanford Research Institute in the 1970s [Nitzan et al., 1977]. Range accuracies approach those of pulsed laser TOF methods. Only a slight advantage is gained over pulsed TOF rangefinding, however, since the time-measurement problem is replaced by the need for fairly sophisticated phase-measurement electronics [Depkovich and Wolfe, 1984]. Because of the limited information obtainable from a single range point, laser-based systems are often scanned in one or more directions by either electromechanical or acousto-optical mechanisms.

4.2.1 Odetics Scanning Laser Imaging System

Odetics, Inc., Anaheim, CA, developed an adaptive and versatile scanning laser rangefinder in the early 1980s for use on *ODEX 1*, a six-legged walking robot [Binger and Harris, 1987; Byrd and DeVries, 1990]. The system determines distance by phase-shift measurement, constructing three-dimensional range pictures by panning and tilting the sensor across the field of view. The phase-shift measurement technique was selected over acoustic-ranging, stereo vision and structured light alternatives because of the inherent accuracy and fast update rate.

The imaging system is broken down into the two major subelements depicted in Figure 4.23: the scan unit and the electronics unit. The scan unit houses the laser source, the photodetector, and the scanning mechanism. The laser source is a GaAlAs laser diode emitting at a wavelength of 820 nanometers; the power output is adjustable under software control between 1 to 50 mW. Detection of the returned energy is achieved through use of an avalanche photodiode whose output is routed to the phase-measuring electronics.

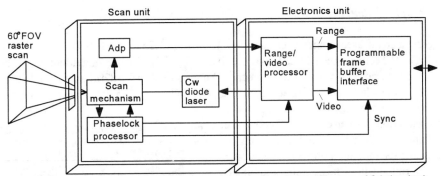

Figure 4.23: Block diagram of the Odetics scanning laser rangefinder. (Courtesy of Odetics, Inc.)

The scanning hardware consists of a rotating polygonal mirror which pans the laser beam across the scene, and a planar mirror whose back-and-forth nodding motion tilts the beam for a realizable field of view of 60 degrees in azimuth and 60 degrees in elevation. The scanning sequence follows a raster-scan pattern and can illuminate and detect an array of 128×128 pixels at a frame rate of 1.2 Hz [Boltinghouse et al., 1990].

The second subelement, the electronics unit, contains the range calculating and video processor as well as a programmable frame buffer interface. The range and video processor is responsible for controlling the laser transmission, activation of the scanning mechanism, detection of the returning energy, and determination of range values. Distance is calculated through a proprietary phase-detection scheme, reported to be fast, fully digital, and self-calibrating with a high signal-to-noise ratio. The minimum observable range is 0.46 meters (1.5 ft), while the maximum range without ambiguity due to phase shifts greater than 360 degrees is 9.3 meters (30 ft).

For each pixel, the processor outputs a range value and a video reflectance value. The video data are equivalent to that obtained from a standard black-and-white television camera, except that interference due to ambient light and shadowing effects are eliminated. The reflectance value is compared to a prespecified threshold to eliminate pixels with insufficient return intensity to be properly processed, thereby eliminating potentially invalid range data; range values are set to maximum for all such pixels [Boltinghouse and Larsen, 1989]. A 3×3 *neighborhood median filter* is used to further filter out noise from data qualification, specular reflection, and impulse response [Larson and Boltinghouse, 1988].

The output format is a 16-bit data word consisting of the range value in either 8 or 9 bits, and the video information in either 8 or 7 bits, respectively. The resulting range resolution for the system is 3.66 centimeters (1.44 in) for the 8-bit format, and 1.83 centimeters (0.72 in) with 9 bits. A buffer interface provides interim storage of the data and can execute single-word or whole-block direct-memory-access transfers to external host controllers under program control. Information can also be routed directly to a host without being held in the buffer. Currently, the interface is designed to support *VAX, VME-Bus, Multibus*, and IBM-*PC/AT* equipment. The scan and electronics unit together weigh 31 lb and require 2 A at 28 VDC.

4.2.2 ESP *Optical Ranging System*

A low-cost near-infrared rangefinder (see Fig. 4.24, Fig. 4.25, and Tab. 4.10) was developed in 1989 by ESP Technologies, Inc., Lawrenceville, NJ [ESP], for use in autonomous robot cart navigation in factories and similar environments. An eyesafe 2 mW, 820-nanometer LED source is 100 percent modulated at 5 MHz and used to form a collimated 2.5 centimeters (1 in) diameter transmit beam that is unconditionally eye-safe. Reflected radiation is focused by a 10-centimeter (4 in) diameter coaxial Fresnel lens onto the photodetector; the measured phase shift is proportional to the round-trip distance to the illuminated object. The *Optical Ranging System (ORS-1)* provides three outputs: range and angle of the target, and an automatic gain control (AGC) signal [Miller and Wagner, 1987]. Range resolution at 6.1 meters (20 ft) is approximately 6 centimeters (2.5 in), while angular resolution is about 2.5 centimeters (1 in) at a range of 1.5 meters (5 ft).

Table 4.10: Selected specifications for the LED-based near-infrared *Optical Ranging System*. (Courtesy of ESP Technologies, Inc.)

Parameter	Value	Units
Accuracy	< 6	in
AGC output	1-5	V
Output power	2	mW
Beam width	2.5	cm
	1	in
Dimensions	15×15×30	cm
	6×6×12	in
Weight		lb
Power	12	VDC
	2	A

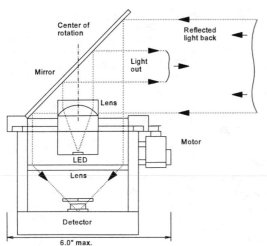

Figure 4.24: Schematic drawing of the *ORS-1* ranging system. (Courtesy of ESP Technologies, Inc.)

Figure 4.25: The *ORS-1* ranging system. (Courtesy of ESP Technologies, Inc.)

The *ORS-1* AGC output signal is inversely proportional to the received signal strength and provides information about a target's near-infrared reflectivity, warning against insufficient or excessive signal return [ESP, 1992]. Usable range results are produced only when the corresponding gain signal is within a predetermined operating range. A rotating mirror mounted at 45 degrees to the optical axis provides 360-degree polar-coordinate coverage. It is driven at 1 to 2 rps by a motor fitted with an integral incremental encoder and an optical indexing sensor that signals the completion of each revolution. The system is capable of simultaneous operation as a wideband optical communication receiver [Miller and Wagner, 1987].

4.2.3 Acuity Research *AccuRange 3000*

Acuity Research, Inc., [ACUITY], Menlo Park, CA, has recently introduced an interesting product capable of acquiring unambiguous range data from 0 to 20 meters (0 to 66 ft) using a proprietary technique similar to conventional phase-shift measurement (see Tab. 4.11). The *AccuRange 3000* (see Figure 4.26) projects a collimated beam of near-infrared or visible laser light, amplitude modulated with a non-sinusoidal waveform at a 50-percent duty cycle. A 63.6-millimeter (2.5 in) collection aperture surrounding the laser diode emitter on the front face of the cylindrical housing gathers any reflected energy returning from the

Figure 4.26: The *AccuRange 3000* distance measuring sensor provides a square-wave output that varies inversely in frequency as a function of range. (Courtesy of Acuity Research, Inc.)

target, and compares it to the outgoing reference signal to produce a square-wave output with a period of oscillation proportional to the measured range. The processing electronics reportedly are substantially different, however, from heterodyne phase-detection systems [Clark, 1994].

The frequency of the output signal varies from approximately 50 MHz at zero range to 4 MHz at 20 meters (66 ft). The distance to
target can be determined through use of a frequency-to-voltage converter, or by measuring the period with a hardware or software timer [Clark, 1994]. Separate 0 to 10 V analog outputs are provided for returned signal amplitude, ambient light, and temperature to facilitate dynamic calibration for optimal accuracy in demanding applications. The range output changes within 250 nanoseconds to reflect any change in target distance, and all outputs are updated within a worst-case time frame of only 3 μs. This rapid response rate (up to 312.5 kHz for all outputs with the optional SCSI interface) allows the beam to be manipulated at a 1,000

Table 4.11: Selected specifications for the *AccuRange 3000* distance measurement sensor. (Courtesy of Acuity Research, Inc.)

Parameter	Value	Units
Laser output	5	mW
Beam divergence	0.5	mrad
Wavelength	780/670	nm
Maximum range	20	m
	65	ft
Minimum range	0	m
Accuracy	2	mm
Sample rate	up to 312.5	kHz
Response time	3	μs
Diameter	7.6	cm
	3	in
Length	14	cm
	5.5	in
Weight	510	g
	18	oz
Power	5 and 12	VDC
	250 and 50	mA

to 2,000 Hz rate with the mechanical-scanner option shown in Figure 4.27. A 45-degree balanced-mirror arrangement is rotated under servo-control to deflect the coaxial outgoing and incoming beams for full 360-degree planar coverage.

It is worthwhile noting that the *AccuRange 3000* appears to be quite popular with commercial and academic lidar developers. For example, TRC (see Sec. 4.2.5 and 6.3.5) is using this sensor in their Lidar and Beacon Navigation products, and the University of Kaiserslautern, Germany, (see Sec. 8.2.3) has used the *AccuRange 3000* in their in-house-made lidars.

Figure 4.27: A 360° beam-deflection capability is provided by an optional single axis rotating scanner. (Courtesy of Acuity Research, Inc.)

4.2.4 TRC Light Direction and Ranging System

Transitions Research Corporation (TRC), Danbury, CT, offers a low-cost *lidar* system (see Figure 4.23) for detecting obstacles in the vicinity of a robot and/or estimating position from local landmarks, based on the previously discussed Acuity Research *AccuRange 3000* unit. TRC adds a 2-DOF scanning mechanism employing a gold front-surfaced mirror specially mounted on a vertical pan axis that rotates between 200 and 900 rpm. The tilt axis of the scanner is mechanically synchronized to nod one complete cycle (down 45° and back to horizontal) per 10 horizontal scans, effectively creating a protective spiral of detection coverage around the robot [TRC, 1994] (see Fig. 4.29). The tilt axis can be mechanically disabled if so desired for 360-degree azimuthal scanning at a fixed elevation angle.

A *68HC11* microprocessor automatically compen sates for variations in ambient lighting and sensor temperature, and reports range, bearing, and elevation data via an Ethernet or RS-232 interface. Power requirements are 500 mA at 12 VDC and 100 mA at 5 VDC. Typical operating parameters are listed in Table 4.12.

Table 4.12: Selected specifications for the TRC *Light Direction and Ranging System.* (Courtesy of Transitions Research Corp.)

Parameter		Value	Units
Maximum range		12	m
		39	ft
Minimum range		0	m
Laser output		6	mW
Wavelength		780	nm
Beam divergence		0.5	mrad
Modulation frequency		2	MHz
Accuracy	(range)	25	mm
		1	in
Resolution	(range)	5	mm
		0.2	in
	(azimuth)	0.18	°
Sample rate		25	kHz
Scan rate		200-900	rpm
Size	(scanner)	13×13×35	cm
		5×5×13.7	in
	(electronics)	30×26×5	cm
		12×10×2	in
Weight		4.4	lb
Power		12 and 5	VDC
		500 and 100	mA

Figure 4.28: The TRC *Light Direction and Ranging System* incorporates a two-axis scanner to provide full-volume coverage sweeping 360° in azimuth and 45° in elevation. (Courtesy of Transitions Research Corp.)

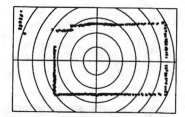 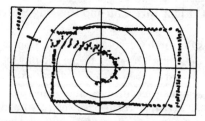

Figure 4.29: LightRanger data plotted from scans of a room. An open door at the upper left and a wall in the corridor detected through the open doorway are seen in the image to the left. On the right a trail has been left by a person walking through the room. (Courtesy of Transitions Research Corp.)

4.2.5 Swiss Federal Institute of Technology's "3-D Imaging Scanner"

Researchers at the Swiss Federal Institute of Technology, Zürich, Switzerland, have developed an optical rangefinder designed to overcome many of the problems associated with commercially available optical rangefinders [Adams, 1995]. The design concepts of the *3-D Imaging Scanner* have been derived from Adam's earlier research work at Oxford University, U.K. [Adams, 1992]. Figure 4.30 shows the working prototype of the sensor. The transmitter consists of an eye-safe high-powered (250 mW) Light Emitting Diode (LED) that provides a range resolution of 4.17 cm/° of phase shift between transmitted and received beams. More detailed specifications are listed in Table 4.13.

The *3-D Imaging Scanner* is now in an advanced prototype stage and the developer plans to market it in the near future [Adams, 1995].

Table 4.13: Preliminary specifications for the *3-D Imaging Scanner*. (Courtesy of [Adams, 1995].)

Parameter		Value	Units
Maximum range		15	m
		50	ft
Minimum range		0	m
LED power (eye-safe)		1	mW
Sweep	(horizontal)	360	°
	(vertical — "nod")	130	°
Resolution	(range)	~20	mm
		0.8	in
	(azimuth)	0.072	°
Sample rate		8	kHz
Size	(diameter×height)	14×27	cm
		5.5×10	in
	(electronics)	Not yet determined	
Weight		Not yet determined	
Power		+12 V @ 400 mA	
		-12 V @ 20 mA	

Figure 4.30: The *3-D Imaging Scanner* consists of a transmitter which illuminates a target and a receiver to detect the returned light. A range estimate from the sensor to the target is then produced. The mechanism shown sweeps the light beam horizontally and vertically. (Courtesy of [Adams, 1995].)

These are some special design features employed in the 3-D Imaging Scanner:

- Each range estimate is accompanied by a range variance estimate, calibrated from the received light intensity. This quantifies the system's confidence in each range data point.

- Direct "crosstalk" has been removed between transmitter and receiver by employing circuit neutralization and correct grounding techniques.

- A software-based discontinuity detector finds spurious points between edges. Such spurious points are caused by the finite optical beamwidth, produced by the sensor's transmitter.

- The newly developed sensor has a tuned load, low-noise, FET input, bipolar amplifier to remove amplitude and ambient light effects.

- Design emphasis on high-frequency issues helps improve the linearity of the amplitude-modulated continuous-wave (phase measuring) sensor.

Figure 4.31 shows a typical scan result from the *3-D Imaging Scanner*. The scan is a pixel plot, where the horizontal axis corresponds to the number of samples recorded in a complete 360-degree rotation of the sensor head, and the vertical axis corresponds to the number of 2-dimensional scans recorded. In Figure 4.31 330 readings were recorded per revolution of the sensor mirror in each horizontal plane, and there were 70 complete revolutions of the mirror. The geometry viewed is "wrap-around geometry," meaning that the vertical pixel set at horizontal coordinate zero is the same as that at horizontal coordinate 330.

4.2.6 Improving Lidar Performance

Unpublished results from [Adams, 1995] show that it is possible to further improve the already good performance of lidar systems. For example, in some commercially available sensors the measured phase shift is not only a function of the sensor-to-target range, but also of the received signal amplitude and ambient light conditions [Vestli et al., 1993]. Adams demonstrates this effect in the sample scan shown in Figure 4.32a. This scan was obtained with the ESP *ORS-1* sensor (see Sec. 4.2.3). The solid lines in Figure 4.32 represent the actual environment and each " × " shows a single range data point. The triangle marks the sensor's position in each case. Note the non-linear behavior of the sensor between points A and B.

Figure 4.32b shows the results from the same ESP sensor, but with the receiver unit redesigned and rebuilt by Adams. Specifically, Adams removed the automatic gain controlled circuit, which is largely responsible for the amplitude-induced range error, and replaced it with four soft limiting amplifiers.

This design approximates the behavior of a logarithmic amplifier. As a result, the weak signals are amplified strongly, while stronger signals remain virtually unamplified. The resulting near-linear signal allows for more accurate phase measurements and hence range determination.

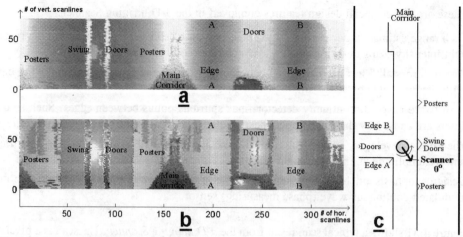

Figure 4.31: Range and intensity scans obtained with Adams' *3-D Imaging Scanner.*
a. In the *range scan* the brightness of each pixel is proportional to the range of the signal received (darker pixels are closer).
b. In the *intensity scan* the brightness of each pixel is proportional to the amplitude of the signal received. (Courtesy of Adams.)

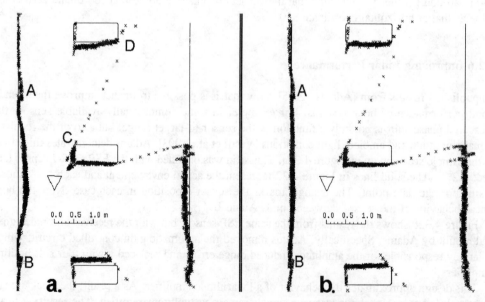

Figure 4.32: Scanning results obtained from the ESP ORS-1 lidar. The triangles represent the sensor's position; the lines represent a simple plan view of the environment and each small cross represents a single range data point.
a. Some non-linearity can be observed for scans of straight surfaces (e.g., between points A and B).
b. Scanning result after applying the signal compression circuit described in [Adams and Probert, 1995].
(Reproduced with permission from [Adams and Probert, 1995].)

Note also the spurious data points between edges (e.g., between C and D). These may be attributed to two potential causes:

- The "*ghost-in-the-machine problem*," in which crosstalk directly between the transmitter and receiver occurs even when no light is returned. Adams' solution involves circuit neutralization and proper grounding procedures.

- The "*beamwidth problem*," which is caused by the finite transmitted width of the light beam. This problem shows itself in form of range points lying between the edges of two objects located at different distances from the lidar. To overcome this problem Adams designed a software filter capable of finding and rejecting erroneous range readings. Figure 4.33 shows the lidar map after applying the software filter.

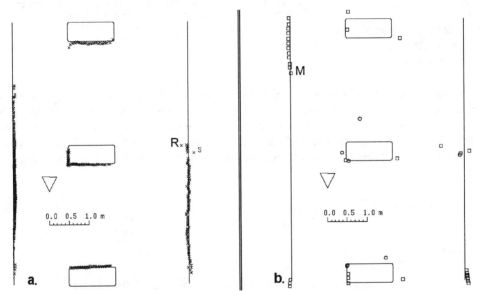

Figure 4.33: Resulting lidar map after applying a software filter.
a. "Good" data that successfully passed the software filter; R and S are "bad" points that "slipped through."
b. Rejected erroneous data points. Point M (and all other square data points) was rejected because the amplitude of the received signal was too low to pass the filter threshold.
(Reproduced with permission from [Adams and Probert, 1995].)

4.3 Frequency Modulation

A closely related alternative to the amplitude-modulated phase-shift-measurement ranging scheme is frequency-modulated (FM) radar. This technique involves transmission of a continuous electromagnetic wave modulated by a periodic triangular signal that adjusts the carrier frequency above and below the mean frequency f_0 as shown in Figure 4.34. The transmitter emits a signal that varies in frequency as a linear function of time:

$$f(t) = f_0 + at \qquad (4.7)$$

where
a = constant
t = elapsed time.

This signal is reflected from a target and arrives at the receiver at time $t + T$.

$$T = \frac{2d}{c} \qquad (4.8)$$

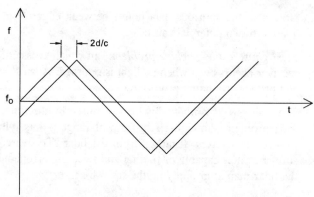

Figure 4.34: The received frequency curve is shifted along the time axis relative to the reference frequency [Everett, 1995].

where
T = round-trip propagation time
d = distance to target
c = speed of light.

The received signal is compared with a reference signal taken directly from the transmitter. The received frequency curve will be displaced along the time axis relative to the reference frequency curve by an amount equal to the time required for wave propagation to the target and back. (There might also be a vertical displacement of the received waveform along the frequency axis, due to the Doppler effect.) These two frequencies when combined in the mixer produce a beat frequency F_b:

$$F_b = f(t) - f(T + t) = aT \qquad (4.9)$$

where
a = constant.

This beat frequency is measured and used to calculate the distance to the object:

$$d = \frac{F_b c}{4 F_r F_d} \qquad (4.10)$$

where
d = range to target
c = speed of light
F_b = beat frequency
F_r = repetition (modulation) frequency
F_d = total FM frequency deviation.

Distance measurement is therefore directly proportional to the difference or beat frequency, and as accurate as the linearity of the frequency variation over the counting interval.

Advances in wavelength control of laser diodes now permit this radar ranging technique to be used with lasers. The frequency or wavelength of a laser diode can be shifted by varying its temperature. Consider an example where the wavelength of an 850-nanometer laser diode is shifted by 0.05 nanometers in four seconds: the corresponding frequency shift is 5.17 MHz per nanosecond. This laser beam, when reflected from a surface 1 meter away, would produce a beat frequency of 34.5 MHz. The linearity of the frequency shift controls the accuracy of the system; a frequency linearity of one part in 1000 yards yields an accuracy of 1 millimeter.

The frequency-modulation approach has an advantage over the phase-shift-measurement technique in that a single distance measurement is not ambiguous. (Recall phase-shift systems must perform two or more measurements at different modulation frequencies to be unambiguous.) However, frequency modulation has several disadvantages associated with the required linearity and repeatability of the frequency ramp, as well as the coherence of the laser beam in optical systems. As a consequence, most commercially available FMCW ranging systems are radar-based, while laser devices tend to favor TOF and phase-detection methods.

4.3.1 Eaton VORAD Vehicle Detection and Driver Alert System

VORAD Technologies [VORAD-1], in joint venture with [VORAD-2], has developed a commercial millimeter-wave FMCW Doppler radar system designed for use on board a motor vehicle [VORAD-1]. The *Vehicle Collision Warning System* employs a 12.7×12.7-centimeter (5×5 in) antenna/transmitter-receiver package mounted on the front grill of a vehicle to monitor speed of and distance to other traffic or obstacles on the road (see Figure 4.35). The flat etched-array antenna radiates approximately 0.5 mW of power at 24.725 GHz directly down the roadway

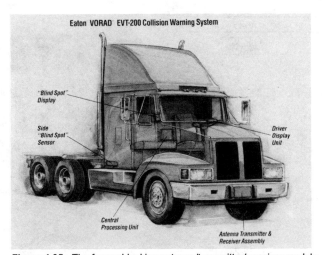

Figure 4.35: The forward-looking antenna/transmitter/ receiver module is mounted on the front of the vehicle at a height between 50 and 125 cm, while an optional side antenna can be installed as shown for blind-spot protection. (Courtesy of VORAD-2).

in a narrow directional beam. A GUNN diode is used for the transmitter, while the receiver employs a balanced-mixer detector [Woll, 1993].

The *Electronics Control Assembly* (see Figure 4.36) located in the passenger compartment or cab can individually distinguish up to 20 moving or stationary objects [Siuru, 1994] out to a maximum range of 106 meters (350 ft); the closest three targets within a prespecified warning distance are tracked at a 30 Hz rate. A Motorola *DSP 56001* and an Intel *87C196* microprocessor calculate range and range-rate information from the RF data and analyze the results in conjunction with vehicle velocity, braking, and steering-angle information. If necessary, the *Control Display Unit* alerts the operator if warranted of potentially hazardous driving situations with a series of caution lights and audible beeps.

As an optional feature, the Vehicle Colli-sion Warning System offers blind-spot detection along the right-hand side of the vehicle out to 4.5 meters (15 ft). The Side Sensor transmitter employs a dielectric resonant oscillator operating in pulsed-Doppler mode at 10.525 GHz, using a flat etched-array antenna with a beamwidth of about 70 degrees [Woll, 1993]. The system microprocessor in the Electronics Control Assembly analyzes the signal strength and frequency components from the Side Sensor subsystem in conjunction with vehicle speed and steering inputs, and activates audible and visual LED alerts if a dangerous condition is thought to exist. (Selected specifications are listed in Tab. 4.14.)

Among other features of interest is a recording feature, which stores 20 minutes of the most recent historical data on a removable EEPROM memory card for post-accident reconstruction. This data includes steering, braking, and idle time. Greyhound Bus Lines recently completed installation of the VORAD radar on all of its 2,400 buses [Bulkeley, 1993], and subsequently reported a 25-year low accident record [Greyhound, 1994]. The entire system weighs just 3 kilograms (6.75 lb), and operates from 12 or 24 VDC with a nominal power consumption of 20 W. An RS-232 digital output is available.

Table 4.14: Selected specifications for the Eaton VORAD *EVT-200 Collision Warning System.* (Courtesy of VORAD-1.)

Parameter	Value	Units
Effective range	0.3-107	m
	1-350	ft
Accuracy	3	%
Update rate	30	Hz
Host platform speed	0.5-120	mph
Closing rate	0.25-100	mph
Operating frequency	24.725	GHz
RF power	0.5	mW
Beamwidth (horizontal)	4	°
(vertical)	5	°
Size (antenna)	15×20×3.8	cm
	6×8×1.5	in
(electronics unit)	20×15×12.	cm
	7	in
	8×6×5	
Weight (total)	6.75	lb
Power	12-24	VDC
	20	W
MTBF	17,000	hr

Figure 4.36: The electronics control assembly of the Vorad *EVT-200 Collision Warning System.* (Courtesy of VORAD-2.)

4.3.2 Safety First Systems Vehicular Obstacle Detection and Warning System

Safety First Systems, Ltd., Plainview, NY, and General Microwave, Amityville, NY, have teamed to develop and market a 10.525 GHz microwave unit (see Figure 4.37) for use as an automotive blind-spot alert for drivers when backing up or changing lanes [Siuru, 1994]. The narrowband (100-kHz) modified-FMCW technique uses patent-pending phase discrimination augmentation for a 20-fold increase in achievable resolution. For example, a conventional FMCW system operating at 10.525 GHz with a 50 MHz bandwidth is limited to a best-case range resolution of approximately 3 meters (10 ft), while the improved approach can resolve distance to within 18 centimeters (0.6 ft) out to 12 meters (40 ft) [SFS]. Even greater accuracy and maximum ranges (i.e., 48 m — 160 ft) are possible with additional signal processing.

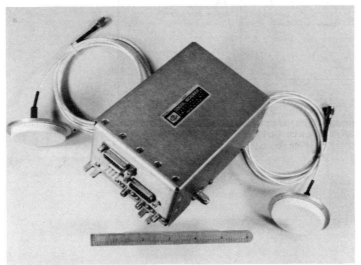

Figure 4.37: Safety First/General Microwave Corporation's Collision Avoidance Radar, Model 1707A with two antennas. (Courtesy of Safety First/General Microwave Corp.)

A prototype of the system delivered to Chrysler Corporation uses conformal bistatic microstrip antennae mounted on the rear side panels and rear bumper of a minivan, and can detect both stationary and moving objects within the coverage patterns shown in Figure 4.38. Coarse range information about reflecting targets is represented in four discrete range bins with individual TTL output lines: 0 to 1.83 meters (0 to 6 ft), 1.83 to 3.35 meters (6 to 11 ft), 3.35 to 6.1 meters (11 to 20 ft), and > 6.1 m (20 ft). Average radiated power is about 50 μW with a three-percent duty cycle, effectively eliminating adjacent-system interference. The system requires 1.5 A from a single 9 to 18 VDC supply.

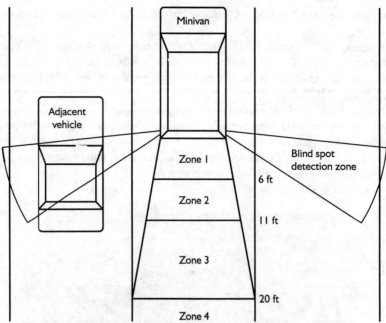

Figure 4.38: The Vehicular Obstacle Detection and Warning System employs a modified
FMCW ranging technique for blind-spot detection when backing up or changing lanes.
(Courtesy of Safety First Systems, Ltd.)

Part II
Systems and Methods for Mobile Robot Positioning

Tech-Team leaders Chuck Cohen, Frank Koss, Mark Huber, and David Kortenkamp (left to right) fine-tune CARMEL in preparation of the 1992 Mobile Robot Competition in San Jose, CA. The efforts paid off: despite its age, CARMEL proved to be the most agile among the contestants, winning first place honors for the University of Michigan.

CHAPTER 5
ODOMETRY AND OTHER DEAD-RECKONING METHODS

Odometry is the most widely used navigation method for mobile robot positioning. It is well known that odometry provides good short-term accuracy, is inexpensive, and allows very high sampling rates. However, the fundamental idea of odometry is the integration of incremental motion information over time, which leads inevitably to the accumulation of errors. Particularly, the accumulation of orientation errors will cause large position errors which increase proportionally with the distance traveled by the robot. Despite these limitations, most researchers agree that odometry is an important part of a robot navigation system and that navigation tasks will be simplified if odometric accuracy can be improved. Odometry is used in almost all mobile robots, for various reasons:

- Odometry data can be fused with absolute position measurements to provide better and more reliable position estimation [Cox, 1991; Hollingum, 1991; Byrne et al., 1992; Chenavier and Crowley, 1992; Evans, 1994].

- Odometry can be used in between absolute position updates with landmarks. Given a required positioning accuracy, increased accuracy in odometry allows for less frequent absolute position updates. As a result, fewer landmarks are needed for a given travel distance.

- Many mapping and landmark matching algorithms (for example: [Gonzalez et al., 1992; Chenavier and Crowley, 1992]) assume that the robot can maintain its position well enough to allow the robot to look for landmarks in a limited area and to match features in that limited area to achieve short processing time and to improve matching correctness [Cox, 1991].

- In some cases, odometry is the only navigation information available; for example: when no external reference is available, when circumstances preclude the placing or selection of landmarks in the environment, or when another sensor subsystem fails to provide usable data.

5.1 Systematic and Non-Systematic Odometry Errors

Odometry is based on simple equations (see Chapt. 1) that are easily implemented and that utilize data from inexpensive incremental wheel encoders. However, odometry is also based on the assumption that wheel revolutions can be translated into linear displacement relative to the floor. This assumption is only of limited validity. One extreme example is wheel slippage: if one wheel was to slip on, say, an oil spill, then the associated encoder would register wheel revolutions even though these revolutions would not correspond to a linear displacement of the wheel.

Along with the extreme case of total slippage, there are several other more subtle reasons for inaccuracies in the translation of wheel encoder readings into linear motion. All of these error sources fit into one of two categories: *systematic errors* and *non-systematic errors*.

Systematic Errors
- Unequal wheel diameters.
- Average of actual wheel diameters differs from nominal wheel diameter.

- Actual wheelbase differs from nominal wheelbase.
- Misalignment of wheels.
- Finite encoder resolution.
- Finite encoder sampling rate.

Non-Systematic Errors
- Travel over uneven floors.
- Travel over unexpected objects on the floor.
- Wheel-slippage due to:
  ○ slippery floors.
  ○ overacceleration.
  ○ fast turning (skidding).
  ○ external forces (interaction with external bodies).
  ○ internal forces (castor wheels).
  ○ non-point wheel contact with the floor.

The clear distinction between systematic and non-systematic errors is of great importance for the effective reduction of odometry errors. For example, systematic errors are particularly grave because they accumulate constantly. On most smooth indoor surfaces systematic errors contribute much more to odometry errors than non-systematic errors. However, on rough surfaces with significant irregularities, non-systematic errors are dominant. The problem with non-systematic errors is that they may appear unexpectedly (for example, when the robot traverses an unexpected object on the ground), and they can cause large position errors. Typically, when a mobile robot system is installed with a hybrid odometry/landmark navigation system, the frequency of the landmarks is determined empirically and is based on the worst-case systematic errors. Such systems are likely to fail when one or more large non-systematic errors occur.

It is noteworthy that many re searchers develop algorithms that estimate the position uncertainty of a dead-reckoning robot (e.g., [Tonouchi et al., 1994; Komoriya and Oyama, 1994].) With this approach each computed robot position is surrounded by a characteristic "error ellipse," which indicates a region of uncertainty for the robot's actual position (see Figure 5.1) [Tonouchi et

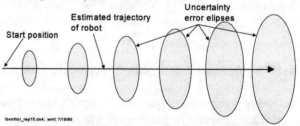

Figure 5.1: Growing "error ellipses" indicate the growing position uncertainty with odometry. (Adapted from [Tonouchi et al., 1994].)

al., 1994; Adams et al., 1994]. Typically, these ellipses grow with travel distance, until an absolute position measurement reduces the growing uncertainty and thereby "resets" the size of the error ellipse. These error estimation techniques must rely on error estimation parameters derived from observations of the vehicle's dead-reckoning performance. Clearly, these parameters can take into account only systematic errors, because the magnitude of non-systematic errors is unpredictable.

5.2   Measurement of Odometry Errors

One important but rarely addressed difficulty in mobile robotics is the *quantitative* measurement of odometry errors. Lack of well-defined measuring procedures for the quantification of odometry errors results in the poor calibration of mobile platforms and incomparable reports on odometric accuracy in scientific communications. To overcome this problem Borenstein and Feng [1995a; 1995c] developed methods for quantitatively measuring systematic odometry errors and, to a limited degree, non-systematic odometry errors. These methods rely on a simplified error model, in which two of the systematic errors are considered to be dominant, namely:

- the error due to unequal wheel diameters, defined as

$$E_d = D_R/D_L \qquad\qquad\qquad\qquad (5.1)$$

   where D_R and D_L are the *actual* wheel diameters of the right and left wheel, respectively.

- The error due to uncertainty about the effective wheelbase, defined as

$$E_b = b_{actual}/b_{nominal} \qquad\qquad\qquad\qquad (5.2)$$

   where b is the wheelbase of the vehicle.

5.2.1 Measurement of Systematic Odometry Errors

To better understand the motivation for Borenstein and Feng's method (discussed in Sec. 5.2.1.2), it will be helpful to investigate a related method first. This related method, described in Section 5.2.1.1, is intuitive and widely used (e.g., [Borenstein and Koren, 1987; Cybermotion, 1988; Komoriya and Oyama, 1994; Russell, 1995], but it is a fundamentally unsuitable benchmark test for differential-drive mobile robots.

5.2.1.1 The Unidirectional Square-Path Test — A Bad Measure for Odometric Accuracy

Figure 5.2a shows a 4×4 meter unidirectional square path. The robot starts out at a position x_0, y_0, θ_0, which is labeled START. The starting area should be located near the corner of two perpendicular walls. The walls serve as a fixed reference before and after the run: measuring the distance between three specific points on the robot and the walls allows accurate determination of the robot's absolute position and orientation.

To conduct the test, the robot must be programmed to traverse the four legs of the square path. The path will return the vehicle to the starting area but, because of odometry and controller errors, not precisely to the starting position. Since this test aims at determining odometry errors and not controller errors, the vehicle does not need to be programmed to return to its starting position precisely — returning approximately to the starting area is sufficient. Upon completion of the square path, the experimenter again measures the absolute position of the vehicle, using the fixed walls as a reference. These absolute measurements are then compared to the position and orientation of the vehicle as computed from odometry data. The result is a set of *return position errors* caused by odometry and denoted ϵx, ϵy, and $\epsilon \theta$.

$$\epsilon x = x_{abs} - x_{calc}$$
$$\epsilon y = y_{abs} - y_{calc} \qquad (5.3)$$
$$\epsilon \theta = \theta_{abs} - \theta_{calc}$$

where

$\epsilon x, \epsilon y, \epsilon \theta$ = position and orientation errors due to odometry

$x_{abs}, y_{abs}, \theta_{abs}$ = absolute position and orientation of the robot

$x_{calc}, y_{calc}, \theta_{calc}$ = position and orientation of the robot as computed from odometry.

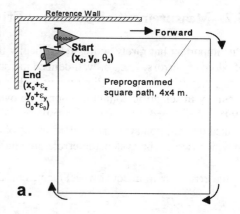

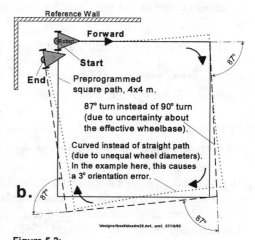

Figure 5.2:
The unidirectional square path experiment.
a. The nominal path.
b. Either one of the two significant errors E_b or E_d can cause the same final position error.

The path shown in Figure 5.2a comprises of four straight-line segments and four pure rotations about the robot's centerpoint, at the corners of the square. The robot's end position shown in Figure 5.2a visualizes the odometry error.

While analyzing the results of this experiment, the experimenter may draw two different conclusions: The odometry error is the result of unequal wheel diameters, E_d, as shown by the slightly curved trajectory in Figure 5.2b (dotted line). Or, the odometry error is the result of uncertainty about the wheelbase, E_b. In the example of Figure 5.2b, E_b caused the robot to turn 87 degrees instead of the desired 90 degrees (dashed trajectory in Figure 5.2b).

As one can see in Figure 5.2b, either one of these two cases *could* yield approximately the same position error. The fact that two different error mechanisms might result in the same overall error may lead an experimenter toward a serious mistake: correcting only one of the two error sources in software. This mistake is so serious because it will yield apparently "excellent" results, as shown in the example in Figure 5.3. In this example, the experimenter began "improving" performance by adjusting the wheelbase b in the control software. According to the dead-reckoning equations for differential-drive vehicles (see Eq. (1.5) in Sec. 1.3.1), the experimenter needs only to increase the value of b to make the robot turn more in each nominal 90-degree turn. In doing so, the experimenter will soon have adjusted b to the seemingly "ideal" value that will cause the robot to turn 93 degrees, thereby effectively compensating for the 3-degree orientation error introduced by each slightly curved (but nominally straight) leg of the square path.

One should note that another popular test path, the "figure-8" path [Tsumura et al., 1981; Borenstein and Koren, 1985; Cox, 1991] can be shown to have the same shortcomings as the unidirectional square path.

5.2.1.2 The Bidirectional Square-Path Experiment

The detailed example of the preceding section illustrates that the unidirectional square path experiment is unsuitable for testing odometry performance in differential-drive platforms, because it can easily conceal two mutually compensating odometry errors. To overcome this problem, Borenstein and Feng [1995a; 1995c] introduced the *bidirectional square-path* experiment, called *University of Michigan Benchmark* (UMBmark). UMBmark requires that the square path experiment be performed in both clockwise and counterclockwise direction. Figure 5.4 shows that the concealed dual error from the example in Figure 5.3 becomes clearly visible when the square path is performed in the opposite direction.

This is so because the two dominant systematic errors, which may compensate for each other when run in only one direction, add up to each other and increase the overall error when run in the opposite direction.

The result of the bidirectional square-path experiment might look similar to the one shown in Figure 5.5, which presents actual experimental results with an off-the-shelf TRC *LabMate* robot [TRC] carrying an evenly distributed load. In this experiment the robot was programmed to follow a 4×4 meter square path, starting at (0,0). The stopping positions for five runs each in clockwise (cw) and counterclockwise (ccw) directions are shown in Figure 5.5. Note that Figure 5.5 is an enlarged view of the target area. The results of Figure 5.5 can be interpreted as follows:

- The stopping positions after cw and ccw runs are clustered in two distinct areas.

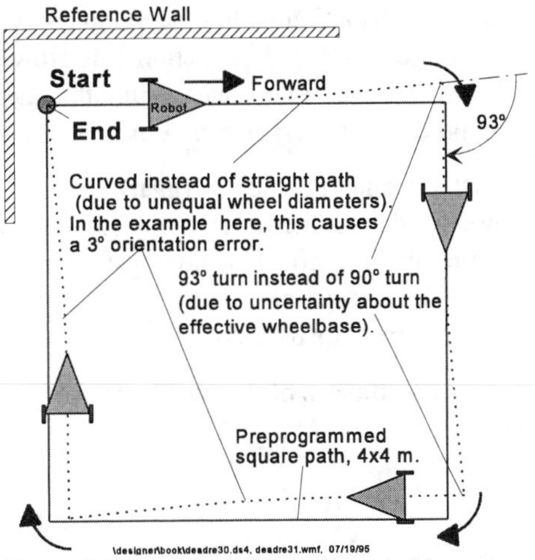

Figure 5.3: The effect of the two dominant systematic dead-reckoning errors E_b and E_d. Note how both errors may cancel each other out when the test is performed in only one direction.

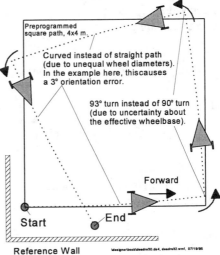

Figure 5.4: The effect of the two dominant systematic odometry errors E_b and E_d: when the square path is performed in the opposite direction one may find that the errors add up.

- pThe distribution within the cw and ccw clusters are the result of non-systematic errors, such as those mentioned in Section 5.1. However, Figure 5.5 shows that in an uncalibrated vehicle, traveling over a reasonably smooth concrete floor, the contribution of *systematic* errors to the total odometry error can be notably larger than the contribution of non-systematic errors.

After conducting the UMBmark experiment, one may wish to derive a single numeric value that expresses the odometric accuracy (with respect to systematic errors) of the tested vehicle. In order to minimize the effect of non-systematic errors, it has been suggested [Komoriya and Oyama, 1994; Borenstein and Feng, 1995c] to consider the center of gravity of each cluster as representative for the systematic odometry errors in the cw and ccw directions.

The coordinates of the two centers of gravity are computed from the results of Equation (5.3) as

$$x_{c.g.,cw/ccw} = \frac{1}{n}\sum_{i=1}^{n} \epsilon x_{i,cw/ccw}$$

$$y_{c.g.,cw/ccw} = \frac{1}{n}\sum_{i=1}^{n} \epsilon y_{i,cw/ccw}$$

(5.4)

where $n = 5$ is the number of runs in each direction.

The absolute offsets of the two centers of gravity from the origin are denoted $r_{c.g.,cw}$ and $r_{c.g.,ccw}$ (see Fig. 5.5) and are given by

$$r_{c.g.,cw} = \sqrt{(x_{c.g.,cw})^2 + (y_{c.g.,cw})^2}$$

(5.5a)

and

$$r_{c.g.,ccw} = \sqrt{(x_{c.g.,ccw})^2 + (y_{c.g.,ccw})^2} \ .$$

(5.5b)

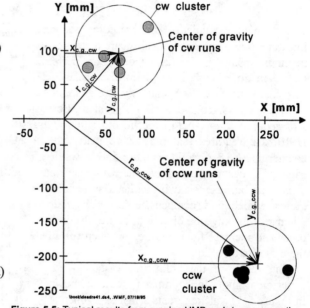

Figure 5.5: Typical results from running UMBmark (a square path run in both cw and ccw directions) with an uncalibrated vehicle.

Finally, the larger value among $r_{c.g.,\ cw}$ and $r_{c.g.,\ ccw}$ is defined as the "*measure of odometric accuracy for systematic errors*":

$$E_{max,syst} = \max(r_{c.g.,cw} \ ; \ r_{c.g.,ccw}) \ .$$

(5.6)

The reason for not using the *average* of the two centers of gravity $r_{c.g.,cw}$ and $r_{c.g.,ccw}$ is that for practical applications one needs to worry about the *largest* possible odometry error. One should also note that the final orientation error $\epsilon\theta$ is not considered explicitly in the expression for $E_{\text{max,syst}}$. This is because all systematic orientation errors are implied by the final position errors. In other words, since the square path has fixed-length sides, systematic orientation errors translate directly into position errors.

5.2.2 Measurement of Non-Systematic Errors

Some limited information about a vehicle's susceptibility to non-systematic errors can be derived from the spread of the return position errors that was shown in Figure 5.5. When running the UMBmark procedure on smooth floors (e.g., a concrete floor without noticeable bumps or cracks), an indication of the magnitude of the non-systematic errors can be obtained from computing the estimated standard deviation σ. However, Borenstein and Feng [1994] caution that there is only limited value to knowing σ, since σ reflects only on the interaction between the vehicle and a certain floor. Furthermore, it can be shown that from comparing σ from two different robots (even if they traveled on the same floor), one cannot necessarily conclude that the robots with the larger σ showed higher susceptibility to non-systematic errors.

In real applications it is imperative that the *largest possible disturbance* be determined and used in testing. For example, the estimated standard deviation of the test in Figure 5.5 gives no indication at all as to what error one should expect if one wheel of the robot inadvertently traversed a large bump or crack in the floor. For the above reasons it is difficult (perhaps impossible) to design a generally applicable quantitative test procedure for non-systematic errors. However, Borenstein [1994] proposed an easily reproducible test that would allow comparing the susceptibility to non-systematic errors of different vehicles. This test, called the *extended UMBmark*, uses the same bidirectional square path as UMBmark but, in addition, introduces artificial bumps. Artificial bumps are introduced by means of a common, round, electrical household-type cable (such as the ones used with 15 A six-outlet power strips). Such a cable has a diameter of about 9 to 10 millimeters. Its rounded shape and plastic coating allow even smaller robots to traverse it without too much physical impact. In the proposed extended UMBmark test the cable is placed 10 times under one of the robot's wheels, during motion. In order to provide better repeatability for this test and to avoid mutually compensating errors, Borenstein and Feng [1994] suggest that these 10 bumps be introduced as evenly as possible. The bumps should also be introduced during the first straight segment of the square path, and always under the wheel that faces the inside of the square. It can be shown [Borenstein, 1994b] that the most noticeable effect of each bump is a fixed orientation error in the direction of the wheel that encountered the bump. In the TRC *LabMate*, for example, the orientation error resulting from a bump of height $h = 10$ mm is roughly $\Delta\theta = 0.44°$ [Borenstein, 1994b].

Borenstein and Feng [1994] proceed to discuss which measurable parameter would be the most useful for expressing the vehicle's susceptibility to non-systematic errors. Consider, for example, Path A and Path B in Figure 5.6. If the 10 bumps required by the *extended UMBmark* test were concentrated at the beginning of the first straight leg (as shown in exaggeration in Path A), then the return position error would be very small. Conversely, if the 10 bumps were concentrated toward the end of the first straight leg (Path B in Figure 5.6), then the return position error would be larger. Because of this sensitivity of the return position errors to the exact location of the bumps it is not a good idea to use the return position error as an indicator for a robot's susceptibility to non-systematic errors. Instead, the return orientation error $\epsilon\theta$ should be used. Although it is more difficult to measure small angles, measurement of $\epsilon\theta$ is a more consistent quantitative indicator for comparing the performance of different robots. Thus, one can measure and express the susceptibility of a vehicle to non-systematic errors in terms of its *average absolute orientation error* defined as

$$\epsilon\theta_{avrg}^{nonsys} = \frac{1}{n}\sum_{i=1}^{n}|\epsilon\theta_{i,cw}^{nonsys} - \epsilon\theta_{avrg,cw}^{sys}| + \frac{1}{n}\sum_{i=1}^{n}|\epsilon\theta_{i,ccw}^{nonsys} - \epsilon\theta_{avrg,ccw}^{sys}| \qquad (5.7)$$

where $n = 5$ is the number of experiments in cw or ccw direction, superscripts "*sys*" and "*nonsys*" indicate a result obtained from either the regular UMBmark test (for systematic errors) or from the extended UMBmark test (for non-systematic errors). Note that Equation (5.7) improves on the accuracy in identifying non-systematic errors by removing the systematic bias of the vehicle, given by

$$\epsilon\theta_{avrg,cw}^{sys} = \frac{1}{n}\sum_{i=1}^{n}\epsilon\theta_{i,cw}^{sys} \qquad (5.8a)$$

and

$$\epsilon\theta_{avrg,ccw}^{sys} = \frac{1}{n}\sum_{i=1}^{n}\epsilon\theta_{i,ccw}^{sys} \qquad (5.8b)$$

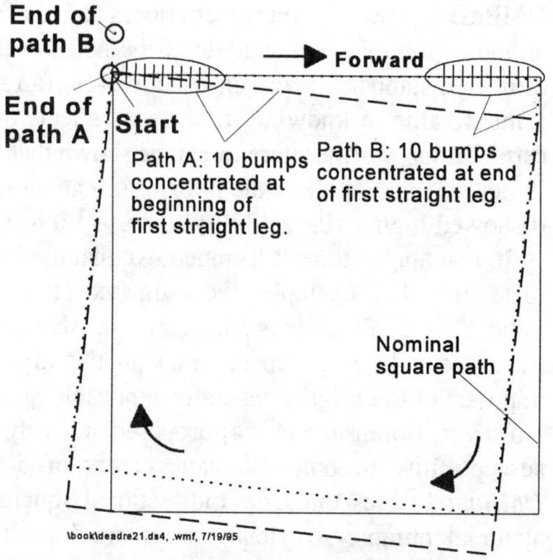

Figure 5.6: The return *position* of the extended UMBmark test is sensitive to the exact location where the 10 bumps were placed. The return *orientation* is not.

Also note that the arguments inside the Sigmas in Equation (5.7) are absolute values of the bias-free return orientation errors. This is because one would want to avoid the case in which two return orientation errors of opposite sign cancel each other out. For example, if in one run $\epsilon\theta = 1°$ and in the next run $\epsilon\theta = -1°$, then one should not conclude that $\epsilon\theta_{avrg}^{nonsys} = 0$. Using the average absolute return error as computed in Equation (5.7) would correctly compute $\epsilon\theta_{avrg}^{nonsys} = 1°$. By contrast, in Equation (5.8) the actual arithmetic average is computed to identify a fixed bias.

5.3   Reduction of Odometry Errors

The accuracy of odometry in commercial mobile platforms depends to some degree on their kinematic design and on certain critical dimensions. Here are some of the design-specific considerations that affect dead-reckoning accuracy:

- Vehicles with a small wheelbase are more prone to orientation errors than vehicles with a larger wheelbase. For example, the differential drive *LabMate* robot from TRC has a relatively small wheelbase of 340 millimeters (13.4 in). As a result, Gourley and Trivedi [1994], suggest that odometry with the *LabMate* be limited to about 10 meters (33 ft), before a new "reset" becomes necessary.

- Vehicles with castor wheels that bear a significant portion of the overall weight are likely to induce slippage when reversing direction (the "shopping cart effect"). Conversely, if the castor wheels bear only a small portion of the overall weight, then slippage will not occur when reversing direction [Borenstein and Koren, 1985].

- It is widely known that, ideally, wheels used for odometry should be "knife-edge" thin and not compressible. The ideal wheel would be made of aluminum with a thin layer of rubber for better traction. In practice, this design is not feasible for all but the most lightweight vehicles, because the odometry wheels are usually also load-bearing drive wheels, which require a somewhat larger ground contact surface.

- Typically the synchro-drive design (see Sec. 1.3.4) provides better odometric accuracy than differential-drive vehicles. This is especially true when traveling over floor irregularities: arbitrary irregularities will affect only one wheel at a time. Thus, since the two other drive wheels stay in contact with the ground, they provide more traction and force the affected wheel to slip. Therefore, overall distance traveled will be reflected properly by the amount of travel indicated by odometry.

Other attempts at improving odometric accuracy are based on more detailed modeling. For example, Larsson et al. [1994] used circular segments to replace the linear segments in each sampling period. The benefits of this approach are relatively small. Boyden and Velinsky [1994] compared (in simulations) conventional odometric techniques, based on kinematics only, to solutions based on the dynamics of the vehicle. They presented simulation results to show that for both differentially and conventionally steered wheeled mobile robots, the kinematic model was accurate only at slower speeds up to 0.3 m/s when performing a tight turn. This result agrees with experimental observations, which suggest that errors due to wheel slippage can be reduced to some degree by limiting the vehicle's speed during turning, and by limiting accelerations.

5.3.1 Reduction of Systematic Odometry Errors

In this section we present specific methods for reducing systematic odometry errors. When applied individually or in combination, these measures can improve odometric accuracy by orders of magnitude.

5.3.1.1 Auxiliary Wheels and Basic Encoder Trailer

It is generally possible to improve odometric accuracy by adding a pair of "knife-edge," non-load-bearing *encoder wheels*, as shown conceptually in Figure 5.7. Since these wheels are not used for transmitting power, they can be made to be very thin and with only a thin layer of rubber as a tire. Such a design is feasible for differential-drive, tricycle-drive, and Ackerman vehicles.

Hongo et al. [1987] had built such a set of encoder wheels, to improve the accuracy of a large differential-drive mobile robot weighing 350 kilograms (770 lb). Hongo et al. report that, after careful calibration, their vehicle had a position error of less than 200 millimeters (8 in) for a travel distance of 50 meters (164 ft). The ground surface on which this experiment was carried out was a "well-paved" road.

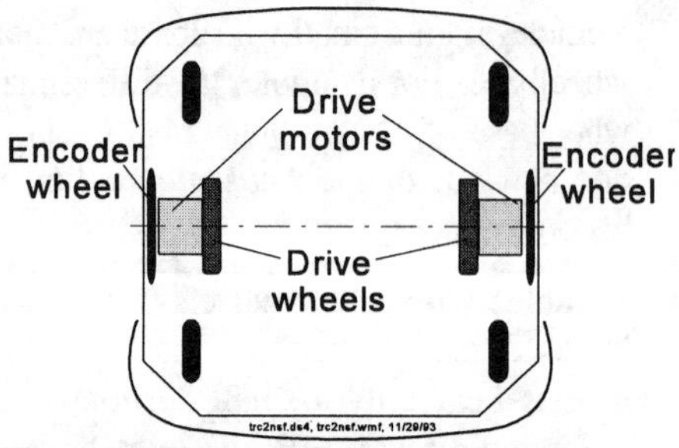

Figure 5.7: Conceptual drawing of a set of *encoder wheels* for a differential drive vehicle.

5.3.1.2 The Basic Encoder Trailer

An alternative approach is the use of a trailer with two encoder wheels [Fan et al., 1994; 1995]. Such an *encoder trailer* was recently built and tested at the University of Michigan (see Figure 5.8). This encoder trailer was designed to be attached to a Remotec *Andros V* tracked vehicle [REMOTEC]. As was explained in Section 1.3, it is virtually impossible to use odometry with tracked vehicles, because of the large amount of slippage between the tracks and the floor during turning. The idea of the encoder trailer is to perform odometry whenever the ground characteristics allow one to do so. Then, when the *Andros* has to move over small obstacles, stairs, or otherwise uneven ground, the encoder trailer would be raised. The argument for this part-time deployment of the encoder trailer is that in many applications the robot may travel *mostly* on reasonably smooth concrete floors and that it would thus benefit *most of the time* from the encoder trailer's odometry.

Figure 5.8: A simple encoder trailer. The trailer here was designed and built at the University of Michigan for use with the Remotec's *Andros V* tracked vehicle. (Courtesy of The University of Michigan.)

5.3.1.3 Systematic Calibration

Another approach to improving odometric accuracy without any additional devices or sensors is based on the careful calibration of a mobile robot. As was explained in Section 5.1, systematic errors are inherent properties of each individual robot. They change very slowly as the result of wear or of different load distributions. Thus, these errors remain almost constant over extended periods of time [Tsumura et al., 1981]. One way to reduce such errors is vehicle-specific calibration. However, calibration is difficult because even minute deviations in the geometry of the vehicle or its parts (e.g., a change in wheel diameter due to a different load distribution) may cause substantial odometry errors.

Borenstein and Feng [1995a; 1995b] have developed a systematic procedure for the measurement and *correction* of odometry errors. This method requires that the UMBmark procedure, described in Section 5.2.1, be run with at least five runs each in cw and ccw direction. Borenstein and Feng define two new error characteristics that are meaningful only in the context of the UMBmark test. These characteristics, called Type A and Type B, represent odometry errors in orientation. A Type A is defined as an orientation error that *reduces (or increases)* the total amount of rotation of the robot during the square-path experiment in *both cw and ccw direction*. By contrast, Type B is defined as an orientation error that *reduces (or increases)* the total amount of rotation of the robot during the square-path experiment in **one direction**, but *increases (or reduces)* the amount of rotation when going in the *other direction*. Examples for Type A and Type B errors are shown in Figure 5.9.

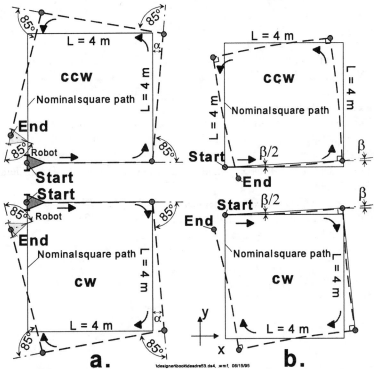

Figure 5.9: Type A and Type B errors in the ccw and cw directions. a. Type A errors are caused only by the wheelbase error E_b. b. Type B errors are caused only by unequal wheel diameters (E_d).

Figure 5.9a shows a case where the robot turned four times for a nominal amount of 90 degrees per turn. However, because the actual wheelbase of the vehicle was larger than the nominal value, the vehicle actually turned only 85 degrees in each corner of the square path. In the example of Figure 5.9 the robot actually turned only $\theta_{total} = 4 \times 85° = 340°$, instead of the desired

$\theta_{\text{nominal}} = 360°$. One can thus observe that in ***both the cw and the ccw*** experiment the robot ends up turning *less* than the desired amount, i.e.,

$$|\theta_{\text{total, cw}}| < |\theta_{\text{nominal}}| \textit{ and } |\theta_{\text{total, ccw}}| < |\theta_{\text{nominal}}| .$$

Hence, the orientation error is of Type A.

In Figure 5.9b the trajectory of a robot with unequal wheel diameters is shown. This error expresses itself in a curved path that adds to the overall orientation at the end of the run in ccw direction, but it reduces the overall rotation in the ccw direction, i.e.,

$$|\theta_{\text{total, ccw}}| > |\theta_{\text{nominal}}| \textit{ but } |\theta_{\text{total,cw}}| < |\theta_{\text{nominal}}| .$$

Thus, the orientation error in Figure 5.9b is of Type B.

In an actual run Type A and Type B errors will of course occur together. The problem is therefore how to distinguish between Type A and Type B errors and how to compute correction factors for these errors from the measured final position errors of the robot in the UMBmark test. This question will be addressed next.

Figure 5.9a shows the contribution of Type A errors. We recall that Type A errors are caused mostly by E_b. We also recall that Type A errors cause too much or too little turning at the corners of the square path. The (unknown) amount of erroneous rotation in each nominal 90-degree turn is denoted as α and measured in [rad].

Figure 5.9b shows the contribution of Type B errors. We recall that Type B errors are caused mostly by the ratio between wheel diameters E_d. We also recall that Type B errors cause a slightly curved path instead of a straight one during the four straight legs of the square path. Because of the curved motion, the robot will have gained an incremental orientation error, denoted β, at the end of each straight leg.

We omit here the derivation of expressions for α and β, which can be found from simple geometric relations in Figure 5.9 (see [Borenstein and Feng, 1995a] for a detailed derivation). Here we just present the results:

$$\alpha = \frac{x_{c.g.,cw} + x_{c.g.,ccw}}{-4L} \frac{180°}{\pi} \tag{5.9}$$

solves for α in [°] and

$$\beta = \frac{x_{c.g.,cw} - x_{c.g.,ccw}}{-4L} \frac{180°}{\pi} \tag{5.10}$$

solves for β in [°].

Using simple geometric relations, the radius of curvature R of the curved path of Figure 5.9b can be found as

$$R = \frac{L/2}{\sin\beta/2} \ . \tag{5.11}$$

Once the radius R is computed, it is easy to determine the ratio between the two wheel diameters that caused the robot to travel on a curved, instead of a straight path

$$E_d = \frac{D_R}{D_L} = \frac{R+b/2}{R-b/2} \ . \tag{5.12}$$

Similarly one can compute the wheelbase error E_b. Since the wheelbase b is directly proportional to the actual amount of rotation, one can use the proportion:

$$\frac{b_{actual}}{90°} = \frac{b_{nominal}}{90° - \alpha} \tag{5.13}$$

so that

$$b_{actual} = \frac{90°}{90° - \alpha} b_{nominal} \tag{5.14}$$

where, per definition of Equation (5.2)

$$E_b = \frac{90°}{90° - \alpha} \ . \tag{5.15}$$

Once E_b and E_d are computed, it is straightforward to use their values as compensation factors in the controller software [see Borenstein and Feng, 1995a; 1995b]. The result is a 10- to 20-fold reduction in systematic errors.

Figure 5.10 shows the result of a typical calibration session. D_R and D_L are the effective wheel diameters, and b is the effective wheelbase.

This calibration procedure can be performed with nothing more than an ordinary tape measure. It takes about two hours to run the complete calibration procedure and measure the individual return errors with a tape measure.

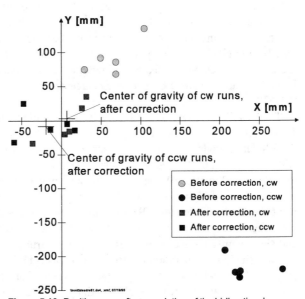

Figure 5.10: Position rrors after completion of the bidirectional square-path experiment (4 x 4 m).
Before calibration: b = 340.00 mm, D_R/D_L = 1.00000.
After calibration: b = 336.17, D_R/D_L = 1.00084.

5.3.2 Reducing Non-Systematic Odometry Errors

This section introduces methods for the reduction of non-systematic odometry errors. The methods discussed in Section 5.3.2.2 may at first confuse the reader because they were implemented on the somewhat complex experimental platform described in Section 1.3.7. However, the methods of Section 5.3.2.2 can be applied to many other kinematic configurations, and efforts in that direction are subject of currently ongoing research at the University of Michigan.

5.3.2.1 Mutual Referencing

Sugiyama [1993] proposed to use two robots that could measure their positions mutually. When one of the robots moves to another place, the other remains still, observes the motion, and determines the first robot's new position. In other words, at any time one robot localizes itself with reference to a fixed object: the standing robot. However, this stop and go approach limits the efficiency of the robots.

5.3.2.2 Internal Position Error Correction

A unique way for reducing odometry errors even further is *Internal Position Error Correction* (IPEC). With this approach two mobile robots mutually correct their odometry errors. However, unlike the approach described in Section 5.3.2.1, the IPEC method works while both robots are in continuous, fast motion [Borenstein, 1994a]. To implement this method, it is required that both robots can measure their relative distance and bearing continuously and accurately. Coincidentally, the MDOF vehicle with compliant linkage (described in Sec. 1.3.7) offers exactly these features, and the IPEC method was therefore implemented and demonstrated on that MDOF vehicle. This implementation is named *Compliant Linkage Autonomous Platform with Position Error Recovery* (CLAPPER).

The CLAPPER's compliant linkage instrumentation was illustrated in Chapter 1, Figure 1.15. This setup provides real-time feedback on the relative position and orientation of the two trucks. An absolute encoder at each end measures the rotation of each truck (with respect to the linkage) with a resolution of 0.3 degrees, while a linear encoder is used to measure the separation distance to within 5 millimeters (0.2 in). Each truck computes its own dead-reckoned position and heading in conventional fashion, based on displacement and velocity information derived from its left and right drive-wheel encoders. By examining the perceived odometry solutions of the two robot platforms in conjunction with their known relative orientations, the *CLAPPER* system can detect and significantly reduce heading errors for both trucks (see video clip in [Borenstein, 1995V].)

The principle of operation is based on the concept of *error growth rate* presented by Borenstein [1994a, 1995], who makes a distinction between "fast-growing" and "slow-growing" odometry errors. For example, when a differentially steered robot traverses a floor irregularity it will immediately experience an appreciable orientation error (i.e., a fast-growing error). The associated lateral displacement error, however, is initially very small (i.e., a slow-growing error), but grows in an unbounded fashion as a consequence of the orientation error. The internal error correction algorithm performs relative position measurements with a sufficiently fast update rate (20 ms) to allow each truck to detect *fast-growing* errors in orientation, while relying on the fact that the lateral position errors accrued by both platforms during the sampling interval were small.

Figure 5.11 explains how this method works. After traversing a bump Truck A's orientation will change (a fact unknown to Truck A's odometry computation). Truck A is therefore expecting to "see" Truck B along the extension of line L_e. However, because of the physically incurred rotation of Truck A, the absolute encoder on truck A will report that truck B is now *actually* seen along line L_m. The angular difference between L_e and L_m is the thus measured odometry orientation error of Truck A, which can be corrected immediately. One should note that even if Truck B encountered a bump at the same time, the resulting rotation of Truck B would not affect the orientation error measurement.

The compliant linkage in essence forms a pseudo-stable heading reference in world coordinates, its own orientation being dictated solely by the relative translations of its end points, which in turn are affected only by the lateral displacements of the two trucks. Since the lateral displacements are *slow growing*, the linkage rotates only a very small amount between encoder samples. The *fast-growing* azimuthal disturbances of the trucks, on the other hand, are not coupled through the rotational joints to the linkage, thus allowing the rotary encoders to detect and quantify the instantaneous orientation errors of the trucks, even when both are in motion. Borenstein [1994a; 1995] provides a more complete description of this innovative concept and reports experimental results indicating improved odometry performance of up to two orders of magnitude over conventional mobile robots.

It should be noted that the rather complex kinematic design of the MDOF vehicle is not necessary to implement the IPEC error correction method. Rather, the MDOF vehicle happened to be available at the time and allowed the University of Michigan researchers to implement and verify the validity of the IPEC approach. Currently, efforts are under way to implement the IPEC method on a tractor-trailer assembly, called "*Smart Encoder Trailer*" (SET), which is shown in Figure 5.12. The principle of operation is illustrated in Figure 5.13. Simulation results, indicating the feasibility of implementing the IPEC method on a tractor-trailer assembly, were presented in [Borenstein, 1994b].

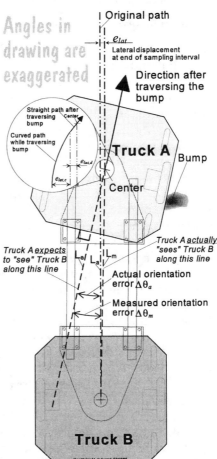

Figure 5.11: After traversing a bump, the resulting change of orientation of Truck A can be measured relative to Truck B.

Figure 5.12: The University of Michigan's "*Smart Encoder Trailer*" (SET) is currently being instrumented to allow the implementation of the IPEC error correction method explained in Section 5.3.2.2. (Courtesy of The University of Michigan.)

5.4   Inertial Navigation

An alternative method for enhancing dead reckoning is inertial navigation, initially developed for deployment on aircraft. The technology was quickly adapted for use on missiles and in outer space, and found its way to maritime usage when the nuclear submarines *Nautilus* and *Skate* were suitably equipped in support of their transpolar voyages in 1958 [Dunlap and Shufeldt, 1972]. The principle of operation involves continuous sensing of minute accelerations in each of the three directional axes and integrating over time to derive velocity and position. A gyroscopically stabilized sensor platform is used to maintain consistent orientation of the three accelerometers throughout this process.

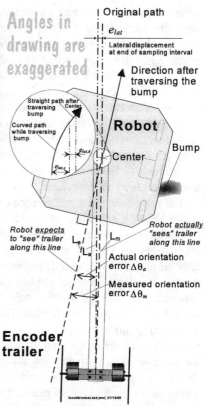

Figure 5.13: Proposed implementation of the IPEC method on a tractor-trailer assembly.

Although fairly simple in concept, the specifics of implementation are rather demanding. This is mainly caused by error sources that adversely affect the stability of the gyros used to ensure correct attitude. The resulting high manufacturing and maintenance costs have effectively precluded any practical application of this technology in the automated guided vehicle industry [Turpin, 1986]. For example, a high-quality *inertial navigation system* (INS) such as would be found in a commercial airliner will have a typical drift of about 1850 meters (1 nautical mile) per hour of operation, and cost between $50K and $70K [Byrne et al., 1992]. High-end INS packages used in ground applications have shown performance of better than 0.1 percent of distance traveled, but cost in the neighborhood of $100K to $200K, while lower performance versions (i.e., one percent of distance traveled) run between $20K to $50K [Dahlin and Krantz, 1988].

Experimental results from the Université Montpellier in France [Vaganay et al., 1993a; 1993b], from the University of Oxford in the U.K. [Barshan and Durrant-Whyte, 1993; 1995], and from the University of Michigan indicate that a purely inertial navigation approach is not realistically advantageous (i.e., too expensive) for mobile robot applications. As a consequence, the use of INS hardware in robotics applications to date has been generally limited to scenarios that aren't readily addressable by more practical alternatives. An example of such a situation is presented by Sammarco [1990; 1994], who reports preliminary results in the case of an INS used to control an autonomous vehicle in a mining application.

Inertial navigation is attractive mainly because it is self-contained and no external motion information is needed for positioning. One important advantage of inertial navigation is its ability to provide fast, low-latency dynamic measurements. Furthermore, inertial navigation sensors typically have noise and error sources that are independent from the external sensors [Parish and Grabbe, 1993]. For example, the noise and error from an inertial navigation system should be quite different from that of, say, a landmark-based system. Inertial navigation sensors are self-contained, non-radiating, and non-jammable. Fundamentally, gyros provide angular rate and accelerometers provide velocity rate information. Dynamic information is provided through direct measurements. However, the main disadvantage is that the angular rate data and the linear velocity rate data must be integrated once and twice (respectively), to provide orientation and linear position, respectively. Thus, even very small errors in the rate information can cause an unbounded growth in the error of integrated measurements. As we remarked in Section 2.2, the price of very accurate laser gyros and optical fiber gyros have come down significantly. With price tags of $1,000 to $5,000, these devices have now become more suitable for many mobile robot applications.

5.4.1 Accelerometers

The suitability of accelerometers for mobile robot positioning was evaluated at the University of Michigan. In this informal study it was found that there is a very poor signal-to-noise ratio at lower accelerations (i.e., during low-speed turns). Accelerometers also suffer from extensive drift, and they are sensitive to uneven grounds, because any disturbance from a perfectly horizontal position will cause the sensor to detect the gravitational acceleration g. One low-cost inertial navigation system aimed at overcoming the latter problem included a tilt sensor [Barshan and Durrant-Whyte, 1993; 1995]. The tilt information provided by the tilt sensor was supplied to the accelerometer to cancel the gravity component projecting on each axis of the accelerometer. Nonetheless, the results obtained from the tilt-compensated system indicate a position drift rate of 1 to 8 cm/s (0.4 to 3.1 in/s), depending on the frequency of acceleration changes. This is an unacceptable error rate for most mobile robot applications.

5.4.2 Gyros

Gyros have long been used in robots to augment the sometimes erroneous dead-reckoning information of mobile robots. As we explained in Chapter 2, mechanical gyros are either inhibitively expensive for mobile robot applications, or they have too much drift. Recent work by Barshan and Durrant-Whyte [1993; 1994; 1995] aimed at developing an INS based on solid-state gyros, and a fiber-optic gyro was tested by Komoriya and Oyama [1994].

5.4.2.1 Barshan and Durrant-Whyte [1993; 1994; 1995]

Barshan and Durrant-Whyte developed a sophisticated INS using two solid-state gyros, a solid-state triaxial accelerometer, and a two-axis tilt sensor. The cost of the complete system was £5,000 (roughly $8,000). Two different gyros were evaluated in this work. One was the ENV-O5S *Gyrostar* from [MURATA], and the other was the Solid State Angular Rate Transducer (*START*) gyroscope manufactured by [GEC]. Barshan and Durrant-Whyte evaluated the performance of these two gyros and found that they suffered relatively large drift, on the order of 5 to 15°/min. The Oxford researchers then developed a sophisticated error model for the gyros, which was subsequently used in an *Extended Kalman Filter* (EKF — see Appendix A). Figure 5.14 shows the results of the experiment for the *START* gyro (left-hand side) and the *Gyrostar* (right-hand side). The thin plotted lines represent the raw output from the gyros, while the thick plotted lines show the output after conditioning the raw data in the EKF.

The two upper plots in Figure 5.14 show the measurement noise of the two gyros while they were stationary (i.e., the rotational rate input was zero, and the gyros should ideally show

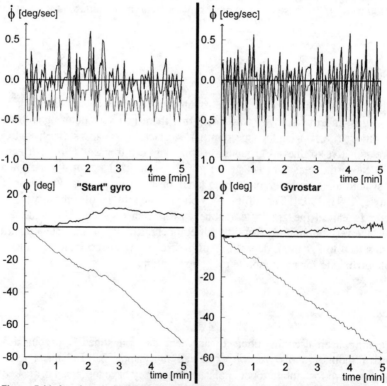

Figure 5.14: Angular rate (top) and orientation (bottom) for zero-input case (i.e., gyro remains stationary) of the *START* gyro (left) and the *Gyrostar* (right) when the bias error is negative. The erroneous observations (due mostly to drift) are shown as the thin line, while the EKF output, which compensates for the error, is shown as the heavy line. (Adapted from [Barshan and Durrant-Whyte, 1995] © IEEE 1995.)

$\dot{\phi}$ = 0 °/s). Barshan and Durrant-Whyte determined that the standard deviation, here used as a measure for the amount of noise, was 0.16°/s for the *START* gyro and 0.24°/s for the *Gyrostar*. The drift in the rate output, 10 minutes after switching on, is rated at 1.35°/s for the *Gyrostar* (drift-rate data for the *START* was not given).

The more interesting result from the experiment in Figure 5.14 is the drift in the angular output, shown in the lower two plots. We recall that in most mobile robot applications one is interested in the heading of the robot, not the rate of change in the heading. The measured rate $\dot{\phi}$ must thus be integrated to obtain ϕ. After integration, any small constant bias in the rate measurement turns into a constant-slope, unbounded error, as shown clearly in the lower two plots of Figure 5.14. At the end of the five-minute experiment, the *START* had accumulated a heading error of -70.8 degrees while that of the *Gyrostar* was -59 degrees (see thin lines in Figure 5.14). However, with the EKF, the accumulated errors were much smaller: 12 degrees was the maximum heading error for the *START* gyro, while that of the *Gyrostar* was -3.8 degrees.

Overall, the results from applying the EKF show a five- to six-fold reduction in the angular measurement after a five-minute test period. However, even with the EKF, a drift rate of 1 to 3°/min can still be expected.

5.4.2.2 Komoriya and Oyama [1994]

Komoriya and Oyama [1994] conducted a study of a system that uses an optical fiber gyroscope, in conjunction with odometry information, to improve the overall accuracy of position estimation. This fusion of information from two different sensor systems is realized through a Kalman filter (see Appendix A).

Figure 5.15 shows a computer simulation of a path-following study without (Figure 5.15a) and with (Figure 5.15b) the fusion of gyro information. The ellipses show the reliability of position estimates (the probability that the robot stays within the ellipses at each estimated position is 90 percent in this simulation).

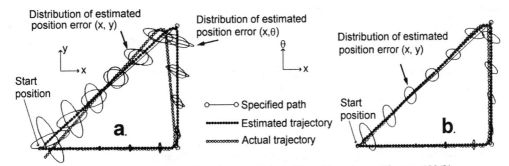

Figure 5.15: Computer simulation of a mobile robot run.. (Adapted from [Komoriya and Oyama, 1994].)
a. Only odometry, without gyro information.   b. Odometry and gyro information fused.

In order to test the effectiveness of their method, Komoriya and Oyama also conducted actual experiments with *Melboy*, the mobile robot shown in Figure 5.16. In one set of experiments *Melboy* was instructed to follow the path shown in Figure 5.17a. *Melboy*'s maximum speed was 0.14 m/s (0.5 ft/s) and that speed was further reduced at the corners of the path in Figure 5.17a. The final position errors without and with gyro information are compared and shown in Figure 5.17b for 20 runs. Figure 5.17b shows that the deviation of the position estimation errors from the mean value is smaller in the case where the gyro data was used (note that a large average deviation from the mean value indicates larger non-systematic errors, as explained in Sec. 5.1). Komoriya and Oyama explain that the noticeable deviation of the mean values from the origin in both cases could be reduced by careful calibration of the systematic errors (see Sec. 5.3) of the mobile robot.

We should note that from the description of this experiment in [Komoriya and Oyama, 1994] it is not immediately evident how the "position estimation error" (i.e., the circles) in Figure 5.17b was found. In our opinion, these points should have been measured by marking the return position of the robot on the floor (or by any

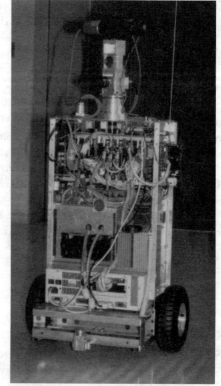

Figure 5.16: *Melboy*, the mobile robot used by Komoriya and Oyama for fusing odometry and gyro data. (Courtesy of [Komoriya and Oyama, 1994].)

equivalent method that records the absolute position of the robot and compares it with the internally computed position estimation). The results of the plot in Figure 5.17b, however, appear to be too accurate for the absolute position error of the robot. In our experience an error on the order of several centimeters, not millimeters, should be expected after completing the path of Figure 5.17a (see, for example, [Borenstein and Koren, 1987; Borenstein and Feng, 1995a; Russel, 1995].) Therefore, we interpret the data in Figure 5.17b as showing a position error that was *computed* by the onboard computer, but not measured absolutely.

5.5 Summary

- Odometry is a central part of almost all mobile robot navigation systems.

- Improvements in odometry techniques will not change their incremental nature, i.e., even for improved odometry, periodic absolute position updates are necessary.

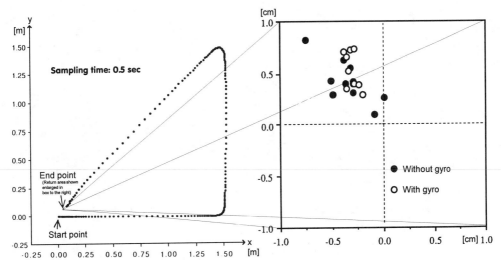

Figure 5.19: Experimental results from *Melboy* using odometry with and without a fiber-optic gyro.
a. Actual trajectory of the robot for a triangular path.
b. Position estimation errors of the robot after completing the path of a. Black circles show the errors without
     gyro; white circles show the errors with the gyro.
(Adapted from [Komoriya and Oyama, 1994].)

- More accurate odometry will reduce the requirements on absolute position updates and will facilitate the solution of landmark and map-based positioning.

- Inertial navigation systems alone are generally inadequate for periods of time that exceed a few minutes. However, inertial navigation can provide accurate short-term information, for example orientation changes during a robot maneuver. Software compensation, usually by means of a Kalman filter, can significantly improve heading measurement accuracy.

CHAPTER 6
ACTIVE BEACON NAVIGATION SYSTEMS

Active beacon navigation systems are the most common navigation aids on ships and airplanes. Active beacons can be detected reliably and provide very accurate positioning information with minimal processing. As a result, this approach allows high sampling rates and yields high reliability, but it does also incur high cost in installation and maintenance. Accurate mounting of beacons is required for accurate positioning. For example, land surveyors' instruments are frequently used to install beacons in a high-accuracy application [Maddox, 1994]. Kleeman [1992] notes that:

> "Although special beacons are at odds with notions of complete robot autonomy in an unstructured environment, they offer advantages of accuracy, simplicity, and speed - factors of interest in industrial and office applications, where the environment can be partially structured."

One can distinguish between two different types of active beacon systems: *trilateration* and *triangulation*.

Trilateration
Trilateration is the determination of a vehicle's position based on distance measurements to known beacon sources. In trilateration navigation systems there are usually three or more transmitters mounted at known locations in the environment and one receiver on board the robot. Conversely, there may be one transmitter on board and the receivers are mounted on the walls. Using time-of-flight information, the system computes the distance between the stationary transmitters and the onboard receiver. *Global Positioning Systems* (GPS), discussed in Section 3.1, are an example of trilateration. Beacon systems based on ultrasonic sensors (see Sec. 6.2, below) are another example.

Triangulation
In this configuration there are three or more active transmitters (usually infrared) mounted at known locations in the environment, as shown in Figure 6.1. A rotating sensor on board the robot registers the angles λ_1, λ_2, and λ_3 at which it "sees" the transmitter beacons relative to the vehicle's longitudinal axis. From these three measurements the unknown x- and y- coordinates and the unknown vehicle orientation θ can be computed. Simple navigation systems of this kind can be built very inexpensively [Borenstein and Koren, 1986]. One problem with this configuration is that the active beacons need to be extremely powerful

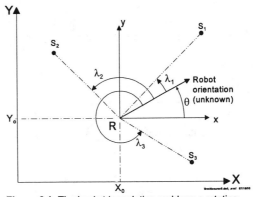

Figure 6.1: The basic triangulation problem: a rotating sensor head measures the three angles λ_1, λ_2, and λ_3 between the vehicle's longitudinal axes and the three sources S_1, S_2, and S_3.

to insure omnidirectional transmission over large distances. Since such powerful beacons are not very practical it is necessary to focus the beacon within a cone-shaped propagation pattern. As a result, beacons are not visible in many areas, a problem that is particularly grave because at least three beacons must be visible for triangulation. A commercially available sensor system based on this configuration (manufactured and marketed by Denning) was tested at the University of Michigan in 1990. The system provided an accuracy of approximately ± 5 centimeters (± 2 in), but the aforementioned limits on the area of application made the system unsuitable for precise navigation in large open areas.

Triangulation methods can further be distinguished by the specifics of their implementation:

a. **Rotating Transmitter-Receiver, Stationary Reflectors** In this implementation there is one rotating laser beam on board the vehicle and three or more stationary retroreflectors are mounted at known locations in the environment.

b. **Rotating Transmitter, Stationary Receivers** Here the transmitter, usually a rotating laser beam, is used on board the vehicle. Three or more stationary receivers are mounted on the walls. The receivers register the incident beam, which may also carry the encoded azimuth of the transmitter.

For either one of the above methods, we will refer to the stationary devices as "*beacons,*" even though they may physically be receivers, retroreflectors, or transponders.

6.1   Discussion on Triangulation Methods

Most of the active beacon positioning systems discussed in Section 6.3 below include computers capable of computing the vehicle's position. One typical algorithm used for this computation is described in [Shoval et al., 1995], but most such algorithms are proprietary because the solutions are non-trivial. In this section we discuss some aspects of triangulation algorithms.

In general, it can be shown that triangulation is sensitive to small angular errors when either the observed angles are small, or when the observation point is on or near a circle which contains the three beacons. Assuming reasonable angular measurement tolerances, it was found that accurate navigation is possible throughout a large area, although error sensitivity is a function of the point of observation and the beacon arrangements [McGillem and Rappaport, 1988].

6.1.1 Three-Point Triangulation

Cohen and Koss [1992] performed a detailed analysis on three-point triangulation algorithms and ran computer simulations to verify the performance of different algorithms. The results are summarized as follows:

- The geometric triangulation method works consistently only when the robot is within the triangle formed by the three beacons. There are areas outside the beacon triangle where the geometric approach works, but these areas are difficult to determine and are highly dependent on how the angles are defined.

- The *Geometric Circle Intersection* method has large errors when the three beacons and the robot all lie on, or close to, the same circle.

- The *Newton-Raphson* method fails when the initial guess of the robot' position and orientation is beyond a certain bound.

- The heading of at least two of the beacons was required to be greater than 90 degrees. The angular separation between any pair of beacons was required to be greater than 45 degrees.

In summary, it appears that none of the above methods alone is always suitable, but an intelligent combination of two or more methods helps overcome the individual weaknesses.

Yet another variation of the triangulation method is the so-called *running fix*, proposed by Case [1986]. The underlying principle of the running fix is that an angle or range obtained from a beacon at time *t*-1 can be utilized at time *t*, as long as the cumulative movement vector recorded since the reading was obtained is added to the position vector of the beacon, thus creating a *virtual* beacon.

6.1.2 Triangulation with More Than Three Landmarks

Betke and Gurvits [1994] developed an algorithm, called the *Position Estimator*, that solves the general triangulation problem. This problem is defined as follows: given the global position of *n* landmarks and corresponding angle measurements, estimate the position of the robot in the global coordinate system. Betke and Gurvits represent the *n* landmarks as complex numbers and formulate the problem as a set of linear equations. By contrast, the traditional law-of-cosines approach yields a set of non-linear equations. Betke and Gurvits also prove mathematically that their algorithm only fails when all landmarks are on a circle or a straight line. The algorithm estimates the robot's position in O(*n*) operations where *n* is the number of landmarks on a two-dimensional map.

Compared to other triangulation methods, the *Position Estimator* algorithm has the following advantages: (1) the problem of determining the robot position in a noisy environment is linearized, (2) the algorithm runs in an amount of time that is a linear function of the number of landmarks, (3) the algorithm provides a position estimate that is close to the actual robot position, and (4) large errors ("outliers") can be found and corrected.

Betke and Gurvits present results of a simulation for the following scenario: the

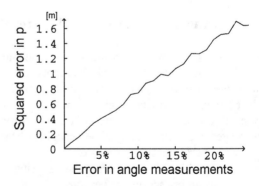

Figure 6.2: Simulation results using the algorithm *Position Estimator* on an input of noisy angle measurements. The squared error in the position estimate *p* (in meters) is shown as a function of measurement errors (in percent of the actual angle). (Reproduced and adapted with permission from [Betke and Gurvits, 1994].)

robot is at the origin of the map, and the landmarks are randomly distributed in a 10×10 meter (32×32 ft) area (see Fig. 6.2). The robot is at the corner of this area. The distance between a landmark and the robot is at most 14.1 meters (46 ft) and the angles are at most 45 degrees. The simulation results show that large errors due to misidentified landmarks and erroneous angle measurements can be found and discarded. Subsequently, the algorithm can be repeated without the outliers, yielding improved results. One example is shown in Figure 6.3, which depicts simulation results using the algorithm *Position Estimator*. The algorithm works on an input of 20 landmarks (not shown in Figure 6.3) that were randomly placed in

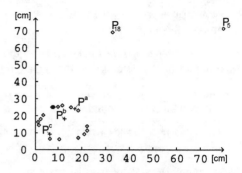

Figure 6.3: Simulation results showing the effect of outliers and the result of removing the outliers. (Reproduced and adapted with permission from [Betke and Gurvits, 1994].)

a 10×10 meters (32×32 ft) workspace. The simulated robot is located at (0, 0). Eighteen of the landmarks were simulated to have a one-percent error in the angle measurement and two of the landmarks were simulated to have a large 10-percent angle measurement error. With the angle measurements from 20 landmarks the *Position Estimator* produces 19 position estimates p_1 - p_{19} (shown as small blobs in Figure 6.3). Averaging these 19 estimates yields the computed robot position. Because of the two landmarks with large angle measurement errors two position estimates are bad: p_5 at (79 cm, 72 cm) and p_{18} at (12.5 cm, 18.3 cm). Because of these poor position estimates, the resulting centroid (average) is at P^a = (17 cm, 24 cm). However, the *Position Estimator* can identify and exclude the two outliers. The centroid calculated without the outliers p_5 and p_{18} is at P^b = (12.5 cm, 18.3 cm). The final position estimate after the *Position Estimator* is applied again on the 18 "good" landmarks (i.e., without the two outliers) is at P^c = (6.5 cm, 6.5 cm).

6.2   Ultrasonic Transponder Trilateration

Ultrasonic trilateration schemes offer a medium- to high-accuracy, low-cost solution to the position location problem for mobile robots. Because of the relatively short range of ultrasound, these systems are suitable for operation in relatively small work areas and only if no significant obstructions are present to interfere with wave propagation. The advantages of a system of this type fall off rapidly, however, in large multi-room facilities due to the significant complexity associated with installing multiple networked beacons throughout the operating area.

Two general implementations exist: 1) a single transducer transmitting from the robot, with multiple fixed-location receivers, and 2) a single receiver listening on the robot, with multiple fixed transmitters serving as beacons. The first of these categories is probably better suited to applications involving only one or at most a very small number of robots, whereas the latter case is basically unaffected by the number of passive receiver platforms involved (i.e., somewhat analogous to the Navstar GPS concept).

6.2.1 IS Robotics 2-D Location System

IS Robotics, Inc. [ISR], Somerville, MA, a spin-off company from MIT's renowned Mobile Robotics Lab, has introduced a beacon system based on an inexpensive ultrasonic trilateration system. This system allows their *Genghis series* robots to localize position to within 12.7 millimeters (0.5 in) over a 9.1×9.1 meter (30×30 ft) operating area [ISR, 1994]. The ISR system consists of a base station master hard-wired to two slave ultrasonic "pingers" positioned a known distance apart (typically 2.28 m — 90 in) along the edge of the operating area as shown in Figure 6.4. Each robot is equipped with a receiving ultrasonic transducer situated beneath a cone-shaped reflector for omnidirectional coverage. Communication between the base station and individual robots is accomplished using a Proxim spread-spectrum (902 to 928 MHz) RF link.

The base station alternately fires the two 40-kHz ultrasonic pingers every half second, each time transmitting a two-byte radio packet in broadcast mode to advise all robots of pulse emission. Elapsed time between radio packet reception and detection of the ultrasonic wave front is used to calculate distance between the robot's current position and the known location of the active beacon. Inter-robot communication is accomplished over the same spread-spectrum channel using a time-division-multiple-access scheme controlled by the base station.

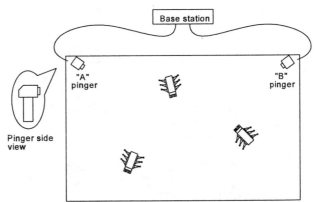

Figure 6.4: The ISR *Genghis* series of legged robots localize x-y position with a master/slave trilateration scheme using two 40 kHz ultrasonic "pingers." (Adapted from [ISR, 1994].)

Principle sources of error include variations in the speed of sound, the finite size of the ultrasonic transducers, non-repetitive propagation delays in the electronics, and ambiguities associated with time-of-arrival detection. The cost for this system is $10,000.

6.2.2 Tulane University 3-D Location System

Researchers at Tulane University in New Orleans, LA, have come up with some interesting methods for significantly improving the time-of-arrival measurement accuracy for ultrasonic transmitter-receiver configurations, as well as compensating for the varying effects of temperature and humidity. In the hybrid scheme illustrated in Figure 6.5, envelope peak detection is employed to establish the approximate time of signal arrival, and to consequently eliminate ambiguity interval problems for a more precise phase-measurement technique that provides final resolution [Figueroa and Lamancusa, 1992]. The desired 0.025 millimeters (0.001 in) range accuracy required a time unit discrimination of 75 nanoseconds at the receiver, which can easily be achieved using fairly simplistic phase measurement circuitry, but only within the interval of a single wavelength. The actual distance from transmitter to receiver is the summation of some integer number of wavelengths (determined by the coarse time-of-arrival measurement) plus that fractional portion of a wavelength represented by the phase measurement results.

Details of this time-of-arrival detection scheme and associated error sources are presented by Figueroa and Lamancusa [1992]. Range measurement accuracy of the prototype system was experimentally determined to be 0.15 millimeters (0.006 in) using both threshold adjustments (based on peak detection) and phase correction, as compared to 0.53 millimeters (0.021 in) for threshold adjustment alone. These high-accuracy requirements were necessary for an application that involved tracking the end-effector of a 6-DOF industrial robot [Figueroa et al, 1992]. The system incorporates seven 90-degree Massa piezoelectric transducers operating at 40 kHz, interfaced to a 33 MHz IBM-compatible PC. The general position-location strategy was based on a trilateration method developed by Figueroa and Mohegan [1994].

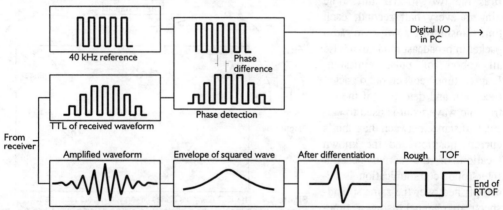

Figure 6.5: A combination of threshold adjusting and phase detection is employed to provide higher accuracy in time-of-arrival measurements in the Tulane University ultrasonic position-location system [Figueroa and Lamancusa, 1992].

The set of equations describing time-of-flight measurements for an ultrasonic pulse propagating from a mobile transmitter located at point (u, v, w) to various receivers fixed in the inertial reference frame can be listed in matrix form as follows [Figueroa and Mohegan, 1994]:

$$\begin{Bmatrix} (t_1 - t_d)^2 \\ (t_2 - t_d)^2 \\ * \\ * \\ * \\ (t_n - t_d)^2 \end{Bmatrix} = \begin{bmatrix} 1 & r_1^2 & 2x_1 & 2y_1 & 2z_1 \\ 1 & r_2^2 & 2x_2 & 2y_2 & 2z_2 \\ * & & & & \\ * & & & & \\ * & & & & \\ 1 & r_n^2 & 2x_n & 2y_n & 2z_n \end{bmatrix} \begin{Bmatrix} \dfrac{p^2}{c^2} \\ \dfrac{1}{c^2} \\ -\dfrac{u}{c^2} \\ -\dfrac{v}{c^2} \\ -\dfrac{w}{c^2} \end{Bmatrix} \tag{6.1}$$

where:

t_i = measured time of flight for transmitted pulse to reach i^{th} receiver
t_d = system throughput delay constant
r_i^2 = sum of squares of ith receiver coordinates
(x_i, y_i, z_i) = location coordinates of ith receiver
(u, v, w) = location coordinates of mobile transmitter
c = speed of sound
p^2 = sum of squares of transmitter coordinates.

The above equation can be solved for the vector on the right to yield an estimated solution for the speed of sound c, transmitter coordinates (u, v, w), and an independent term p^2 that can be compared to the sum of the squares of the transmitter coordinates as a checksum indicator [Figueroa and Mahajan, 1994]. An important feature of this representation is the use of an additional receiver (and associated equation) to enable treatment of the speed of sound itself as an unknown, thus ensuring continuous on-the-fly recalibration to account for temperature and humidity effects. (The system throughput delay constant t_d can also be determined automatically from a pair of equations for $1/c^2$ using two known transmitter positions. This procedure yields two equations with t_d and c as unknowns, assuming c remains constant during the procedure.) A minimum of five receivers is required for an unambiguous three-dimensional position solution, but more can be employed to achieve higher accuracy using a least-squares estimation approach. Care must be taken in the placement of receivers to avoid singularities as defined by Mahajan [1992].

Figueroa and Mahajan [1994] report a follow-up version intended for mobile robot positioning that achieves 0.25 millimeters (0.01 in) accuracy with an update rate of 100 Hz. The prototype system tracks a TRC *LabMate* over a 2.7×3.7 meter (9×12 ft) operating area with five ceiling-mounted receivers and can be extended to larger floor plans with the addition of more receiver sets. An RF link will be used to provide timing information to the receivers and to transmit the subsequent x-y position solution back to the robot. Three problem areas are being further investigated to increase the effective coverage and improve resolution:

- Actual transmission range does not match the advertised operating range for the ultrasonic transducers, probably due to a resonant frequency mismatch between the transducers and electronic circuitry.
- The resolution of the clocks (6 MHz) used to measure time of flight is insufficient for automatic compensation for variations in the speed of sound.
- The phase-detection range-measurement correction sometimes fails when there is more than one wavelength of uncertainty. This problem can likely be solved using the frequency division scheme described by Figueroa and Barbieri [1991].

6.3   Optical Positioning Systems

Optical positioning systems typically involve some type of scanning mechanism operating in conjunction with fixed-location references strategically placed at predefined locations within the operating environment. A number of variations on this theme are seen in practice [Everett, 1995]:

- Scanning detectors with fixed active beacon emitters.
- Scanning emitter/detectors with passive retroreflective targets.
- Scanning emitter/detectors with active transponder targets.
- Rotating emitters with fixed detector targets.

One of the principal problems associated with optical beacon systems, aside from the obvious requirement to modify the environment, is the need to preserve a clear line of sight between the robot and the beacon. Preserving an unobstructed view is sometimes difficult if not impossible in certain applications such as congested warehouse environments. In the case of passive retro-reflective targets, problems can sometimes arise from unwanted returns from other reflective surfaces in the surrounding environment, but a number of techniques exists for minimizing such interference.

6.3.1 Cybermotion Docking Beacon

The automated docking system used on the Cybermotion *Navmaster* robot incorporates the unique combination of a structured-light beacon (to establish bearing) along with a one-way ultrasonic ranging system (to determine standoff distance). The optical portion consists of a pair of near-infrared transceiver units, one mounted on the front of the robot and the other situated in a known position and orientation within the operating environment. These two optical transceivers are capable of full-duplex data transfer between the robot and the dock at a rate of 9600 bits per second. Separate modulation frequencies of 154 and 205 kHz are employed for the uplink and downlink respectively to eliminate crosstalk. Under normal circumstances, the dock-mounted transceiver waits passively until interrogated by an active transmission from the robot. If the interrogation is specifically addressed to the assigned ID number for that particular dock, the dock control computer activates the beacon transmitter for 20 seconds. (Dock IDs are jumper selectable at time of installation.)

Figure 6.6 shows the fixed-location beacon illuminating a 90-degree field of regard broken up into two uniquely identified zones, designated for purposes of illustration here as the *Left Zone* and *Right Zone*. An array of LED emitters in the beacon head is divided by a double-sided mirror arranged along the optical axis and a pair of lenses. Positive zone identification is initiated upon request from the robot in the form of a NAV Interrogation byte transmitted over the optical datalink. LEDs on opposite

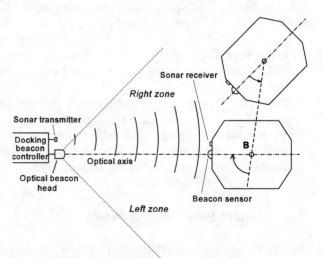

Figure 6.6: The structured-light near-infrared beacon on the Cybermotion battery recharging station defines an optimal path of approach for the *K2A Navmaster* robot [Everett, 1995].

sides of the mirror respond to this *NAV Interrogation* with slightly different coded responses. The robot can thus determine its relative location with respect to the optical axis of the beacon based on the response bit pattern detected by the onboard receiver circuitry.

Once the beacon starts emitting, the robot turns in the appropriate direction and executes the steepest possible (i.e., without losing sight of the beacon) intercept angle with the beacon optical axis. Crossing the optical axis at point B is flagged by a sudden change in the bit pattern of the *NAV Response Byte*, whereupon the robot turns inward to face the dock. The beacon optical axis establishes the nominal path of approach and in conjunction with range offset information uniquely defines the robot's absolute location. This situation is somewhat analogous to a TACAN station [Dodington, 1989] but with a single defined radial.

The offset distance from vehicle to dock is determined in rather elegant fashion by a dedicated non-reflective ultrasonic ranging configuration. This high-frequency (> 200 kHz) narrow-beam (15°) sonar system consists of a piezoelectric transmitter mounted on the docking beacon head and a complimentary receiving transducer mounted on the front of the vehicle. A ranging operation is initiated upon receipt of the *NAV Interrogation Byte* from the robot; the answering *NAV Response Byte* from the docking beacon signals the simultaneous transmission of an ultrasonic pulse. The difference at the robot end between time of arrival for the *NAV Response Byte* over the optical link and subsequent ultrasonic pulse detection is used to calculate separation distance. This dual-transducer master/slave technique assures an unambiguous range determination between two well defined points and is unaffected by any projections on or around the docking beacon and/or face of the robot.

During transmission of a *NAV Interrogation Byte*, the left and right sides of the LED array located on the robot are also driven with uniquely identifiable bit patterns. This feature allows the docking beacon computer to determine the robot's actual heading with respect to the nominal path of approach. Recall the docking beacon's structured bit pattern establishes (in similar fashion) the side of the vehicle centerline on which the docking beacon is located. This heading information is subsequently encoded into the *NAV Response Byte* and passed to the robot to facilitate course correction. The robot closes on the beacon, halting at the defined stop range (not to exceed 8 ft) as repeatedly measured by the docking sonar. Special instructions in the path program can then be used to reset vehicle heading and/or position.

6.3.2 *Hilare*

Early work incorporating passive beacon tracking at the *Laboratoire d'Automatique et d'Analyse des Systemes*, Toulouse, France, involved the development of a navigation subsystem for the mobile robot *Hilare* [Banzil et al., 1981]. The system consisted of two near-infrared emitter/detectors mounted with a 25 centimeters (10 in) vertical separation on a rotating mast, used in conjunction with passive reflective beacon arrays at known locations in three corners of the room.

Each of these beacon arrays was constructed of retroreflective tape applied to three vertical cylinders, which were then placed in a recognizable configuration as shown in Figure 6.7. One of the arrays was inverted so as to be uniquely distinguishable for purposes of establishing an origin. The cylinders were vertically spaced to intersect the two planes of light generated by the rotating optical axes of the two emitters on the robot's mast. A detected reflection pattern as in

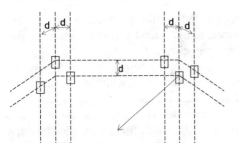

Figure 6.7: Retroreflective beacon array configuration used on the mobile robot *Hilare*. (Adapted from [Banzil et al, 1981].)

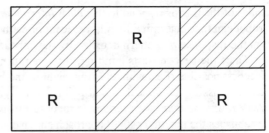

Figure 6.8: A confirmed reflection pattern as depicted above was required to eliminate potential interference from other highly specular surfaces [Banzil et al., 1981].

Figure 6.8 confirmed beacon acquisition. Angular orientation relative to each of the retroreflective arrays was inferred from the stepper-motor commands that drove the scanning mechanism; lateral position was determined through simple triangulation.

6.3.3 NAMCO *LASERNET*

The NAMCO *LASERNET* beacon tracking system (Figure 6.9) employs retroreflective targets distributed throughout the operating area of an automated guided vehicle (AGV) in order to measure range and angular position (Figure 6.10). A servo-controlled rotating mirror pans a near-infrared laser beam through a horizontal arc of 90 degrees at a 20 Hz update rate. When the beam sweeps across a target of known dimensions, a return signal of finite duration is sensed by the detector. Since the targets are all the same size, the signal generated by a close target will be of longer duration than that from a distant one.

Angle measurement is initiated when the scanner begins its sweep from right to left; the laser strikes an internal synchronization photodetector that starts a timing sequence. The beam is then panned across the scene until returned by a retroreflective target in the field of view. The reflected signal is detected by the sensor, terminating the timing sequence (Fig. 6.11). The elapsed time is used to calculate the angular position of the target in the equation [NAMCO, 1989]

$$\theta = Vt_b - 45°$$ (6.2)

where
θ = target angle
V = scan velocity (7,200°/s)
T_b = time between scan initiation and target detection.

Figure 6.9: The *LASERNET* beacon tracking system. (Courtesy of Namco Controls Corp.)

This angle calculation determines either the leading edge of the target, the trailing edge of the target, or the center of the target, depending upon the option selected within the *LASERNET* software option list. The angular accuracy is ±1 percent, and the angular resolution is 0.1 degrees for the analog output; accuracy is within ±.05 percent with a resolution of 0.006 degrees when the RS-232 serial port is used. The analog output is a voltage ranging from 0 to 10 V over the range of -45 to +45 degrees, whereas the RS-232 serial port reports a proportional "count value" from 0 to 15360 over this same range. The system costs $3,400 in its basic configuration, but it has only a limited range of 15 meters (50 ft).

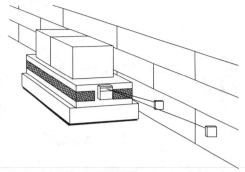

Figure 6.10: The LASERNET system can be used with projecting wall-mounted targets to guide an AGV at a predetermined offset distance. (Courtesy of NAMCO Controls.)

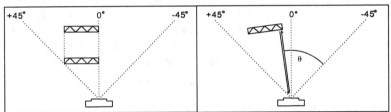

Figure 6.11: a. The perceived width of a retroreflective target of known size is used to calculate range; b. while the elapsed time between sweep initiation and leading edge detection yields target bearing. (Courtesy of NAMCO Controls).

6.3.4 Denning Branch International Robotics *LaserNav Position Sensor*

Denning Branch International Robotics [DBIR], Pittsburgh, PA, offers a laser-based scanning beacon system that computes vehicle position and heading out to 183 meters (600 ft) using cooperative electronic transponders, called *active targets*. A range of 30.5 meters (100 ft) is achieved with simple reflectors (passive targets). The *LaserNav Intelligent Absolute Positioning Sensor*, shown in Figures 6.12 and 6.13, is a non-ranging triangulation system with an absolute bearing accuracy of 0.03 degrees at a scan rate of 600 rpm. The fan-shaped beam is spread 4 degrees vertically to ensure target detection at long range while traversing irregular floor surfaces, with horizontal divergence limited to 0.017 degrees. Each target can be uniquely coded so that the *LaserNav* can distinguish between up to 32 separate active or passive targets during a single scan. The vehicle's x-y position is calculated every 100 milliseconds. The sensor package weighs 4.4 kilograms (10 lb), measures 38 centimeters (15 in) high and 30 centimeters (12 in) in diameter, and has a power consumption of only 300 mA at 12 V. The eye-safe near-infrared laser generates a 1 mW output at a wavelength of 810 nanometers.

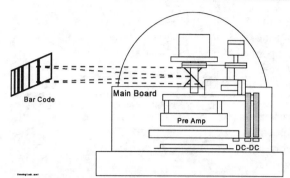

Figure 6.12: Schematics of the Denning Branch International Robotics *LaserNav* laser-based scanning beacon system. (Courtesy of Denning Branch International Robotics.)

Figure 6.13: Denning Branch International Robotics (DBIR) can see *active targets* at up to 183 meters (600 ft) away. It can identify up to 32 active or passive targets. (Courtesy of Denning Branch International Robotics.)

One potential source of problems with this device is the relatively small vertical divergence of the beam: ±2 degrees. Another problem mentioned by the developer [Maddox, 1994] is that *"the LaserNav sensor … is subject to rare spikes of wrong data."* This undesirable phenomenon is likely due to reflections off shiny surfaces other than the passive reflectors. This problem affects probably all light-based beacon navigation systems to some degree. Another source of erroneous beacon readings is bright sunlight entering the workspace through wall openings.

6.3.5 TRC Beacon Navigation System

Transitions Research Corporation [TRC], Danbury, CT, has incorporated their LED-based *LightRanger*, discussed in Section 4.2, into a compact, low-cost navigational referencing system for open-area autonomous platform control. The TRC *Beacon Navigation System* calculates vehicle position and heading at ranges up to 24.4 meters (80 ft) within a quadrilateral area defined by four passive retroreflective beacons [TRC, 1994] (see Figure 6.14). A static 15-second unobstructed view of all four beacons is required for initial acquisition and setup, after which only two beacons must remain in view as the robot moves about. At this time there is no provision to periodically acquire new beacons along a continuous route, so operation is currently constrained to a single zone roughly the size of a small building (i.e., 24.4×24.4 m or 80×80 ft).

System resolution is 120 millimeters (4¾ in) in range and 0.125 degrees in bearing for full 360-degree coverage in the horizontal plane. The scan unit (less processing electronics) is a cube approximately 100 millimeters (4 in) on a

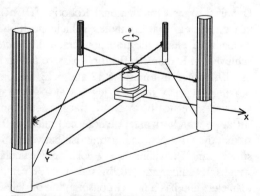

Figure 6.14: The TRC *Beacon Navigation System* calculates position and heading based on ranges and bearings to two of four passive beacons defining a quadrilateral operating area. (Courtesy of TRC.)

side, with a maximum 1-Hz update rate dictated by the 60-rpm scan speed. A dedicated 68HC11 microprocessor continuously outputs navigational parameters (x,y,θ) to the vehicle's onboard controller via an RS-232 serial port. Power requirements are 0.5 A at 12 VDC and 0.1 A at 5 VDC. The system costs $11,000.

6.3.6 Siman Sensors and Intelligent Machines Ltd., *ROBOSENSE*

The *ROBOSENSE* is an eye-safe, scanning laser rangefinder developed by Siman Sensors & Intelligent Machines Ltd., Misgav, Israel (see Figure 6.15). The scanner illuminates retroreflective targets mounted on walls in the environment. It sweeps 360-degree segments in continuous rotation but supplies navigation data even while observing targets in narrower segments (e.g., 180°). The system's output are x- and y-coordinates in a global coordinate system, as well as heading and a confidence level. According to the manufacturer [Siman, 1995], the system is designed to operate under severe or adverse conditions, such as the partial occlusion of the reflectors. A rugged case houses the electro-optical sensor, the navigation computer, the communication module, and the power supply. *ROBOSENSE* incorporates a unique self-mapping feature that does away with the need for precise measurement of the targets, which is needed with other systems.

The measurement range of the *ROBOSENSE* system is 0.3 to 30 meters (1 to 100 ft). The position accuracy is 20 millimeters (3/4 in) and the accuracy in determining the orientation is better than 0.17 degrees. The system can communicate with an onboard computer via serial link, and it updates the position and heading information at a rate of 10 to 40 Hz. *ROBOSENSE* navigates through areas that can be much larger than the system's range. This is done by dividing the whole site map into partial frames, and positioning the system within each frame in the global coordinate system. This method, called *Rolling Frames*, enables *ROBOSENSE* to cover practically unlimited area.

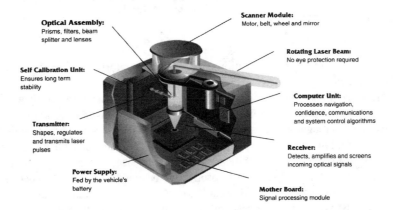

Figure 6.15: The *ROBOSENSE* scanning laser rangefinder was developed by Siman Sensors & Intelligent Machines Ltd., Misgav, Israel. The system determines its own heading and absolute position with an accuracy of 0.17° and 20 millimeters (3/4 in), respectively. (Courtesy of Siman Sensors & Intelligent Machines.)

The power consumption of the *ROBOSENSE* system is less than 20 W at 24 VDC. The price for a single unit is $12,800 and $7,630 each for an order of three units.

6.3.7 Imperial College Beacon Navigation System

Premi and Besant [1983] of the Imperial College of Science and Technology, London, England, describe an AGV guidance system that incorporates a vehicle-mounted laser beam rotating in a horizontal plane that intersects three fixed-location reference sensors as shown in Figure 6.16. The photoelectric sensors are arranged in collinear fashion with equal separation and are individually wired to a common FM transmitter via appropriate electronics so that the time of arrival of laser energy is relayed to a companion receiver on board the vehicle. A digitally coded identifier in the data stream identifies the activated sensor that triggered the transmission, thus allowing the onboard computer to measure the separation angles α_1 and α_2.

AGV position P(x,y) is given by the equations [Premi and Besant, 1983]

$$x = x_1 + r\cos\theta$$
$$y = y_1 + r\sin\theta \qquad (6.3)$$

where

$$r = \frac{a\sin(\alpha_1 + \beta)}{\sin\alpha_1} \qquad (6.4)$$

$$\beta = \arctan\frac{2\tan\alpha_1\tan\alpha_2}{\tan\alpha_2 - \tan\alpha_1} - 1 \qquad (6.5)$$

$$\theta = \phi - \beta \qquad (6.6)$$

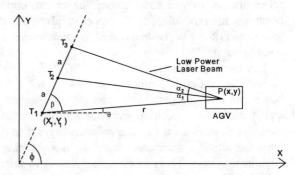

Figure 6.16: Three equidistant collinear photosensors are employed in lieu of retroreflective beacons in the Imperial College laser triangulation system for AGV guidance. (Adapted from [Premi and Besant, 1983].)

An absolute or indexed incremental position encoder that monitors laser scan azimuth is used to establish platform heading.

This technique has some inherent advantages over the use of passive retroreflective targets, in that false acquisition of reflective surfaces is eliminated, and longer ranges are possible since target reflectivity is no longer a factor. More robust performance is achieved through elimination of target dependencies, allowing a more rapid scan rate to facilitate faster positional updates. The one-way nature of the optical signal significantly reduces the size, weight, and cost of the onboard scanner with respect to that required for retroreflective beacon acquisition. Tradeoffs, however, include the increased cost associated with installation of power and communications lines and the need for significantly more expensive beacons. This can be a serious drawback in very-large-area installations, or scenarios where multiple beacons must be incorporated to overcome line-of-sight limitations.

6.3.8 MTI Research CONAC™

A similar type system using a predefined network of fixed-location detectors is currently being built and marketed by MTI Research, Inc., Chelmsford, MA [MTI]. MTI's *Computerized Opto-electronic Navigation and Control*[1] (CONAC) is a relatively low-cost, high-performance navigational referencing system employing a vehicle-mounted laser unit called *STRuctured Opto-electronic - Acquisition Beacon* (STROAB), as shown in Figure 6.17. The scanning laser beam is spread vertically to eliminate critical alignment, allowing the receivers, called *Networked Opto-electronic Acquisition Datums* (NOADs) (see Figure 6.18), to be mounted at arbitrary heights (as illustrated in Figure 6.19). Detection of incident illumination by a NOAD triggers a response over the network to a host PC, which in turn calculates the implied angles α_1 and α_2. An index sensor

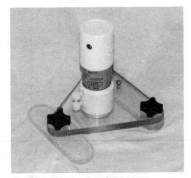

Figure 6.17: A single STROAB beams a vertically spread laser signal while rotating at 3,000 rpm. (Courtesy of, MTI Research Inc.)

built into the STROAB generates a special rotation reference pulse to facilitate heading measurement. Indoor accuracy is on the order of centimeters or millimeters, and better than 0.1 degrees for heading.

The reference NOADs are strategically installed at known locations throughout the area of interest, and daisy chained together with ordinary four-conductor modular telephone cable. Alternatively the NOADS can be radio linked to eliminate cable installation problems, as long as power is independently available to the various NOAD sites. STROAB acquisition range is sufficient to where three NOADS can effectively cover an area of 33,000 m² (over 8 acres) assuming no interfering structures block the view. Additional NOADS are typically employed to increase fault tolerance and minimize ambiguities when two or more robots are operating in close proximity. The optimal set of three NOADS is dynamically selected by the host PC, based on the current location of the robot and any predefined visual barriers. The selected NOADS are individually addressed over the network in accordance with assigned codes (set into DIP switches on the back of each device at time of installation).

An interesting and unconventional aspect of CONAC™ is that no fall-back dead-reckoning capability is incorporated into the system [MacLeod and Chiarella, 1993]. The 3,000 rpm angular rotation speed of the laser STROAB facilitates rapid position updates at a 25 Hz rate, which MTI claims is sufficient for safe automated transit at highway speeds, provided line-of-sight contact is preserved with at least three fixed NOADS. To minimize chances of

Figure 6.18: Stationary NOADs are located at known positions; at least two NOADs are networked and connected to a PC. (Courtesy of MTI Research, Inc.)

---

[1] CONAC is a trademark of MTI.

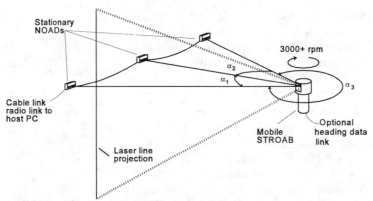

Figure 6.19: The Computerized Opto-electronic Navigation and Control (CONAC™) system employs an onboard, rapidly rotating and vertically spread laser beam, which sequentially contacts the networked detectors. (Courtesy of MTI Research, Inc.)

occlusion, the lightweight (less than 250 g — 9 oz) STROAB is generally mounted as high as possible on a supporting mast.

The ability of the CONAC™ system was demonstrated in an intriguing experiment with a small, radio-controlled race car called *Scooter*. During this experiment, the *Scooter* achieved speeds greater than 6.1 m/s (20 ft/s) as shown by the *Scooters* mid-air acrobatics in Figure 6.20. The small vehicle was equipped with a STROAB and programmed to race along the race course shown in Figure 6.21. The small boxes in Figure 6.21 represent the desired path, while the continuous line represents the position of the vehicle during a typical run. 2,200 data points were collected along the 200 meter (650 ft) long path. The docking maneuver at the end of the path brought the robot to within 2 centimeters (0.8 in) of the desired position. On the tight turns, the *Scooter* decelerated to smoothly execute the hairpin turns.

Figure 6.20: MTI's *Scooter* zips through a race course; tight close-loop control is maintained even in mid-air and at speeds of up to 6.1 m/s (20 ft/s).

CONAC™ Fixed Beacon System

A stationary active beacon
system that tracks an omnidirectional sensor mounted
on the robot is currently
being sold to allow for
tracking multiple units.
(The original CONAC™
system allows only one
beacon to be tracked at a
given time.) The basic system consists of two syn-
chronized stationary bea-
cons that provide bearings
to the mobile sensor to establish its x-y location. A
hybrid version of this approach employs two lasers
in one of the beacons, as

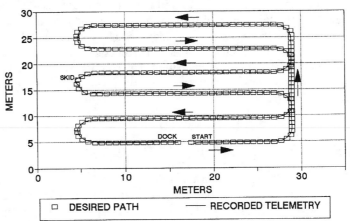

Figure 6.21: Preprogrammed race course and recorded telemetry of the Scooter experiment. Total length: 200 m (650 ft); 2200 data points collected. (Courtesy of MTI Research, Inc.)

illustrated in Figure 6.22, with the lower laser plane tilted from the vertical to provide coverage
along the z-axis for three-dimensional applications. A complete two-dimensional indoor system
is shown in Figure 6.23.

Long-range exterior position accuracy is specified as ± 1.3 millimeters (± 0.5 in) and the
heading accuracy as ± 0.05 degrees. The nominal maximum line-of-sight distance is 250 meters
(780 ft), but larger distances can be covered with a more complex system. The system was

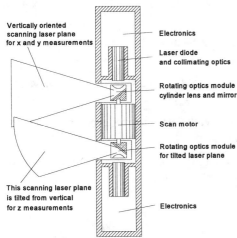

Figure 6.22: Simplified cross section view of the dual-laser position-location system now under development for tracking multiple mobile sensors in 3-D applications. (Courtesy of MTI Research, Inc.)

Figure 6.23: MTI's basic 2-D indoor package. A mobile position transponder (shown in lower center) detects the passing laser emissions generated by the two spread-out stationary laser beacons. (Courtesy of MTI Research, Inc.)

successfully demonstrated in an outdoor environment when MacLeod engineers outfitted a DodgeCaravan with electric actuators for steering, throttle, and brakes, then drove the unmanned vehicle at speeds up to 80 km/h (50 mph) [Baker, 1993]. MTI recently demonstrated the same vehicle at 108 km/h (65 mph). Absolute position and heading accuracies were sufficient to allow the Caravan to maneuver among parked vehicles and into a parking place using a simple AutoCad representation of the environment. Position computations are updated at a rate of 20 Hz. This system represents the current state-of-the-art in terms of active beacon positioning [Fox, 1993; Baker, 1993; Gunther, 1994]. A basic system with one STROAB and three NOADs costs on the order of $4,000.

6.3.9 Lawnmower *CALMAN*

Larsson et al. [1994] from the University of Lulea, Sweden, have converted a large riding lawnmower to fully autonomous operation. This system, called *CALMAN*, uses an onboard rotating laser scanner to illuminate strategically placed vertical retroreflector stripes. These reflectors are attached to tree stems or vertical poles in the environment. Larsson et al. report experimental results from running the vehicle in a parking lot. According to these results, the vehicle had a positioning error of less than 2 centimeters (3/4 in) at speeds of up to 0.3 milliseconds (1 ft/s). The motion of the vehicle was stable at speeds of up to 1 m/s (3.3 ft/s) and became unstable at 1.5 m/s (5 ft/s).

6.4 Summary

We summarize the general characteristics of active beacon systems as follows:

- The environment needs to be modified, and some systems require electric outlets or battery maintenance for stationary beacons.

- A line of sight between transmitter and detector needs to be maintained, i.e., there must be at least two or three *visible* landmarks in the environment.

- Triangulation-based methods are subject to the limitations of triangulation as discussed by Cohen and Koss [1992].

- Active beacon systems have been proven in practice, and there are several commercial systems available using laser, infrared, and ultrasonic transducers.

- In practice, active beacon systems are the choice when high accuracy and high reliability are required.

CHAPTER 7
LANDMARK NAVIGATION

Landmarks are distinct features that a robot can recognize from its sensory input. Landmarks can be geometric shapes (e.g., rectangles, lines, circles), and they may include additional information (e.g., in the form of bar-codes). In general, landmarks have a fixed and known position, relative to which a robot can localize itself. Landmarks are carefully chosen to be easy to identify; for example, there must be sufficient contrast to the background. Before a robot can use landmarks for navigation, the characteristics of the landmarks must be known and stored in the robot's memory. The main task in localization is then to recognize the landmarks reliably and to calculate the robot's position.

In order to simplify the problem of landmark acquisition it is often assumed that the current robot position and orientation are known approximately, so that the robot only needs to look for landmarks in a limited area. For this reason good odometry accuracy is a prerequisite for successful landmark detection.

The general procedure for performing landmark-based positioning is shown in Figure 7.1. Some approaches fall between landmark and map-based positioning (see Chap. 8). They use sensors to sense the environment and then extract distinct structures that serve as landmarks for navigation in the future. These approaches will be discussed in the chapter on map-based positioning techniques.

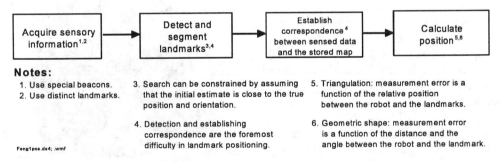

Notes:

1. Use special beacons.
2. Use distinct landmarks.

3. Search can be constrained by assuming that the initial estimate is close to the true position and orientation.

4. Detection and establishing correspondence are the foremost difficulty in landmark positioning.

5. Triangulation: measurement error is a function of the relative position between the robot and the landmarks.

6. Geometric shape: measurement error is a function of the distance and the angle between the robot and the landmark.

Feng1pos.ds4; .wmf

Figure 7.1: General procedure for landmark-based positioning.

Our discussion in this chapter addresses two types of landmarks: "artificial" and "natural." It is important to bear in mind that "natural" landmarks work best in highly structured environments such as corridors, manufacturing floors, or hospitals. Indeed, one may argue that "natural" landmarks work best when they are actually man-made (as is the case in highly structured environments). For this reason, we shall define the terms "natural landmarks" and "artificial landmarks" as follows: *natural landmarks* are those objects or features that are already in the environment and have a function other than robot navigation; *artificial landmarks* are specially designed objects or markers that need to be placed in the environment with the sole purpose of enabling robot navigation.

7.1 Natural Landmarks

The main problem in natural landmark navigation is to detect and match characteristic features from sensory inputs. The sensor of choice for this task is computer vision. Most computer vision-based natural landmarks are long vertical edges, such as doors and wall junctions, and ceiling lights. However, computer vision is an area that is too large and too diverse for the scope of this book. For this reason we will present below only one example of computer vision-based landmark detection, but without going into great detail.

When range sensors are used for natural landmark navigation, distinct signatures, such as those of a corner or an edge, or of long straight walls, are good feature candidates. The selection of features is important since it will determine the complexity in feature description, detection, and matching. Proper selection of features will also reduce the chances for ambiguity and increase positioning accuracy. A natural landmark positioning system generally has the following basic components:

- A sensor (usually computer vision) for detecting landmarks and contrasting them against their background.
- A method for matching observed features with a map of known landmarks.
- A method of computing location and localization errors from the matches.

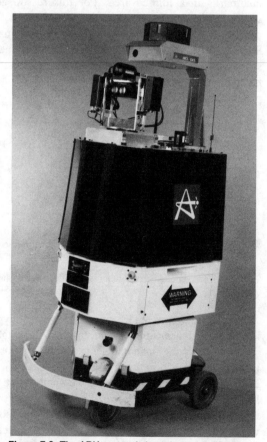

One system that uses natural landmarks has recently been developed in Canada. This project aimed at developing a sophisticated robot system called the *"Autonomous Robot for a Known Environment"* (ARK). The project was carried out jointly by the Atomic Energy of Canada Ltd (AECL) and Ontario Hydro Technologies with support from the University of Toronto and York University [Jenkin et al., 1993]. A Cybermotion K2A+ platform serves as the carrier for a number of sensor subsystems (see Figure 7.2).

Of interest for the discussion here is the ARK navigation module (shown in Figure 7.3). This unit consists of a custom-made pan-and-tilt table, a CCD camera, and an eye-safe IR spot laser rangefinder. Two VME-based cards, a single-board computer, and a microcontroller, provide processing power. The navigation module is used to periodically correct the robot's accumulating odometry errors. The system uses *natural* landmarks

Figure 7.2: The ARK system is based on a modified Cybermotion K2A+. It is one of the few working navigation systems based on natural landmark detection. (Courtesy of Atomic Energy of Canada Ltd.)

such as alphanumeric signs, semi-permanent structures, or doorways. The only criteria used is that the landmark be distinguishable from the background scene by color or contrast.

The ARK navigation module uses an interesting hybrid approach: the system stores (learns) landmarks by generating a three-dimensional "grey-level surface" from a single training image obtained from the CCD camera. A coarse, registered range scan of the same field of view is performed by the laser rangefinder, giving depths for each pixel in the grey-level surface. Both procedures are performed from a known robot position. Later, during operation, when the robot is at an approximately known (from odometry) position within a couple of meters from the training position, the vision system searches for those landmarks that are expected to be visible from the robot's momentary position. Once a

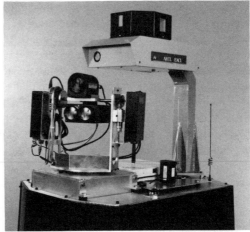

Figure 7.3: AECL's natural landmark navigation system uses a CCD camera in combination with a time-of-flight laser rangefinder to identify landmarks and to measure the distance between landmark and robot. (Courtesy of Atomic Energy of Canada Ltd.)

suitable landmark is found, the projected appearance of the landmark is computed. This *expected* appearance is then used in a coarse-to-fine normalized correlation-based matching algorithm that yields the robot's relative distance and bearing with regard to that landmark. With this procedure the ARK can identify different natural landmarks and measure its position relative to the landmarks.

To update the robot's odometry position the system must find a pair of natural landmarks of known position. Positioning accuracy depends on the geometry of the robot and the landmarks but is typically within a few centimeters. It is possible to pass the robot through standard 90-centimeter (35 in) doorway openings using only the navigation module if corrections are made using the upper corners of the door frame just prior to passage.

7.2   Artificial Landmarks

Detection is much easier with artificial landmarks [Atiya and Hager, 1993], which are designed for optimal contrast. In addition, the exact size and shape of artificial landmarks are known in advance. Size and shape can yield a wealth of geometric information when transformed under the perspective projection.

Researchers have used different kinds of patterns or marks, and the geometry of the method and the associated techniques for position estimation vary accordingly [Talluri and Aggarwal, 1993]. Many artificial landmark positioning systems are based on computer vision. We will not discuss these systems in detail, but we will mention some of the typical landmarks used with computer vision.

Fukui [1981] used a diamond-shaped landmark and applied a least-squares method to find line segments in the image plane. Borenstein [1987] used a black rectangle with four white dots in the corners. Kabuka and Arenas [1987] used a half-white and half-black circle with a unique bar-code for each landmark. Magee and Aggarwal [1984] used a sphere with horizontal and vertical calibration circles to achieve three-dimensional localization from a single image. Other systems use reflective material patterns and strobed light to ease the segmentation and parameter extraction [Lapin, 1992; Mesaki and Masuda, 1992]. There are also systems that use active (i.e., LED) patterns to achieve the same effect [Fleury and Baron, 1992].

The accuracy achieved by the above methods depends on the accuracy with which the geometric parameters of the landmark images are extracted from the image plane, which in turn depends on the relative position and angle between the robot and the landmark. In general, the accuracy decreases with the increase in relative distance. Normally there is a range of relative angles in which good accuracy can be achieved, while accuracy drops significantly once the relative angle moves out of the "good" region.

There is also a variety of landmarks used in conjunction with non-vision sensors. Most often used are bar-coded reflectors for laser scanners. For example, currently ongoing work by Everett on the *Mobile Detection Assessment and Response System* (MDARS) [DeCorte, 1994] uses retro-reflectors, and so does the commercially available system from Caterpillar on their *Self-Guided Vehicle* [Gould, 1990]. The shape of these landmarks is usually unimportant. By contrast, a unique approach taken by Feng et al. [1992] used a circular landmark and applied an optical Hough transform to extract the parameters of the ellipse on the image plane in real time.

7.2.1 Global Vision

Yet another approach is the so-called *global vision* that refers to the use of cameras placed at fixed locations in a workspace to extend the local sensing available on board each AGV [Kay and Luo, 1993]. Figure 7.4 shows a block diagram of the processing functions for vehicle control using global vision.

In global vision methods, characteristic points forming a pattern on the mobile robot are identified and localized from a single view. A probabilistic method is used to select the most probable matching according to geometric characteristics of those percepts. From this reduced search space a prediction-verification loop is applied to identify and to localize the points of the pattern [Fleury and Baron, 1992]. One advantage of this approach is that it allows the operator to monitor robot operation at the same time.

7.3 Artificial Landmark Navigation Systems

Many systems use retroreflective barcodes as artificial landmarks, similar to the ones used in beacon navigation systems. However, in this book we distinguish between retroreflective bar-codes used as artificial landmarks and retroreflective poles used as "beacons." The reason is that if retroreflective markers (with or without bar-code) are attached to the walls of a room and their function is merely to aid in determining the location of the wall, then these markers do not

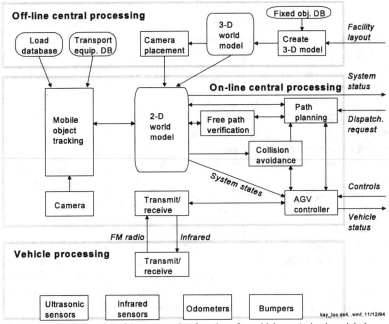

Figure 7.4: Block diagram of the processing functions for vehicle control using global vision. (Adapted from [Kay and Luo, 1993].)

function as beacons. By contrast, if markers are used on arbitrarily placed poles (even if the location of these poles is carefully surveyed), then they act as beacons. A related distinction is the method used for computing the vehicle's position: if triangulation is used, then the reflectors act as beacons.

7.3.1 MDARS Lateral-Post Sensor

Currently ongoing work by Everett on the *Mobile Detection Assessment and Response System* (MDARS) [Everett et al., 1994; DeCorte, 1994] uses passive reflectors in conjunction with a pair of fixed-orientation sensors on board the robot. This technique, called *lateral-post detection*, was incorporated on MDARS to significantly reduce costs by exploiting the forward motion of the robot for scanning purposes. Short vertical strips of 2.5 centimeters (1 in) retroreflective tape are placed on various immobile objects (usually structural-support posts) on either side of a virtual path segment. The exact x-y locations of these tape markers are encoded into the virtual path program. Installation takes only seconds, and since the flat tape does not protrude into the aisle at all, there is little chance of damage from a passing fork truck.

A pair of Banner Q85VR3LP retroreflective proximity sensors mounted on the turret of the *Navmaster* robot face outward to either side as shown in Figure 7.5 These inexpensive sensors respond to reflections from the tape markers along the edges of the route, triggering a "snapshot" *virtual path* instruction that records the current side-sonar range values. The longitudinal position of the platform is updated to the known marker coordinate, while lateral position is inferred from the sonar data, assuming both conditions fall within specified tolerances.

The accuracy of the marker correction is much higher (and therefore assigned greater credibility) than that of the lateral sonar readings due to the markedly different uncertainties associated with the respective targets. The polarized Banner sensor responds only to the presence of a retroreflector while ignoring even highly specular surrounding surfaces, whereas the ultrasonic energy from the sonar will echo back from any

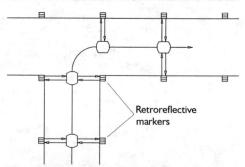

Figure 7.5: Polarized retroreflective proximity sensors are used to locate vertical strips of retroreflective tape attached to shelving support posts in the Camp Elliott warehouse installation of the MDARS security robot [Everett et al, 1994].

reflective surface encountered by its relatively wide beam. Protruding objects in the vicinity of the tape (quite common in a warehouse environment) result in a shorter measured range value than the reference distance for the marker itself. The overall effect on x-y bias is somewhat averaged out in the long run, as each time the vehicle executes a 90-degree course change the association of x- and y-components with tape versus sonar updates is interchanged.

7.3.2 Caterpillar *Self Guided Vehicle*

Caterpillar Industrial, Inc., Mentor, OH, manufactures a free-ranging AGV for materials handling that relies on a scanning laser triangulation scheme to provide positional updates to the vehicle's onboard odometry system. The Class-I laser rotates at 2 rpm to illuminate passive retroreflective bar-code targets affixed to walls or support columns at known locations up to 15 meters (50 ft) away [Gould, 1990; Byrne et al., 1992]. The bar-codes serve to positively identify the reference target and eliminate ambiguities due to false returns from other specular

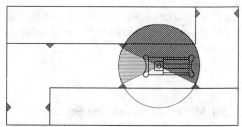

Figure 7.6: Retroreflective bar-code targets spaced 10 to 15 meters (33 to 49 ft) apart are used by the Caterpillar SGV to triangulate position. (Adapted from [Caterpillar, 1991a].)

surfaces within the operating area. An onboard computer calculates x-y position updates through simple triangulation to null out accumulated odometry errors (see Figure 7.6).

Some target occlusion problems have been experienced in exterior applications where there is heavy fog, as would be expected, and minor difficulties have been encountered as well during periods when the sun was low on the horizon [Byrne, 1993]. Caterpillar's *Self Guided Vehicle* (SGV) relies on dead reckoning under such conditions to reliably continue its route for distances of up to 10 meters (33 ft) before the next valid fix.

The robot platform is a hybrid combination of tricycle and differential drive, employing two independent series-wound DC motors powering 45-centimeter (18 in) rear wheels through sealed gear-boxes [CATERPILLAR, 1991]. High-resolution resolvers attached to the single front wheel continuously monitor steering angle and distance traveled. A pair of mechanically scanned near-infrared proximity sensors sweeps the path in front of the vehicle for potential obstructions. Additional near infrared sensors monitor the area to either side of the vehicle, while ultrasonic sensors cover the back.

7.3.3 Komatsu Ltd, *Z-shaped landmark*

Komatsu Ltd. in Tokyo, Japan, is a manufacturer of construction machines. One of Komatsu's research projects aims at developing an unmanned dump truck. As early as 1984, researchers at Komatsu Ltd. developed an unmanned electric car that could follow a previously taught path around the company's premises. The vehicle had two onboard computers, a directional gyrocompass, two incremental encoders on the wheels, and a metal sensor which detected special landmarks along the planned path (see Figure 7.7).

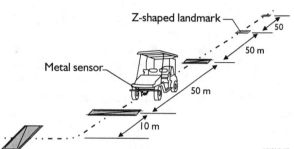

Figure 7.7: Komatsu's Z-shaped landmarks are located at 50-meter (164 ft) intervals along the planned path of the autonomous vehicle. (Courtesy of [Matsuda and Yoshikawa, 1989].)

The accuracy of the vehicle's dead-reckoning system (gyrocompass and encoders) was approximately two percent on the paved road and during straight-line motion only. The mechanical gyrocompass was originally designed for deep-sea fishing boats and its static direction accuracy was 1 degree. On rough terrain the vehicle's dead-reckoning error deteriorated notably. For example, running over a 40-millimeter (1.5 in) height bump and subsequently traveling along a straight line for 50 meters (164 ft), the vehicle's positioning error was 1.4 m (55 in). However, with the Z-shaped landmarks used in this project for periodic recalibration the positioning could

be recalibrated to an accuracy of 10 centimeters (4 in). The 3 meter (118 in) wide landmark was made of 50 millimeter (2 in) wide aluminum strips sandwiched between two rubber sheets. In order to distinguish between "legitimate" metal markings of the landmark and between arbitrary metal objects, additional parallel line segments were used (see Figure 7.8). The metal markers used as landmarks in this experiment are resilient to contamination even in harsh environments. Water, dust, and lighting condition do not affect the readability of the metal sensor [Matsuda and Yoshikawa, 1989].

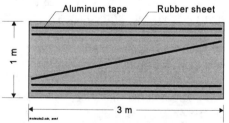

Figure 7.8: The Z-shaped landmark. Note the secondary lines parallel to the horizontal Z-stripes. The secondary lines help distinguish the marker from random metal parts on the road. (Courtesy of [Matsuda and Yoshikawa, 1989].)

Each Z-shaped landmark comprises three line segments. The first and third line segments are in parallel, and the second one is located diagonally between the parallel lines (see Figure7.9). During operation, a metal sensor located underneath the autonomous vehicle detects the three crossing points P_1, P_2, and P_3. The distances, L_1 and L_2, are measured by the incremetal encoders using odometry. After traversing the Z-shaped landmark, the vehicle's lateral deviation X_2 at point P_2 can be computed from

$$X_2 = W(\frac{L_1}{L_1+L_2} - \frac{1}{2}) \qquad (7.1)$$

where X_2 is the lateral position error at point P_2 based on odometry.

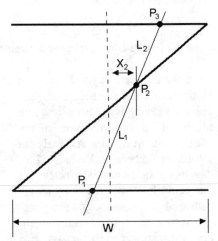

The lateral position error can be corrected after passing through the third crossing point P_3. Note that for this correction method the exact location of the landmark along the line of travel does not have to be known. However, if the location of the landmark is known, then the vehicle's actual position at P_2 can be calculated easily [Matsuda et al., 1989].

The size of the Z-shaped landmark can be varied, according to the expected lateral error of the vehicle. Larger landmarks can be buried under the surface of paved roads for unmanned cars. Smaller landmarks can be installed under factory floor coating or under office carpet. Komatsu has developed such smaller Z-shaped landmarks for indoor robots and AGVs.

Figure 7.9: The geometry of the Z-shaped landmark lends itself to easy and unambiguous computation of the lateral position error X_2. (Courtesy of [Matsuda and Yoshikawa, 1989].)

7.4   Line Navigation

Another type of landmark navigation that has been widely used in industry is line navigation. Line navigation can be thought of as a continuous landmark, although in most cases the sensor used in this system needs to be very close to the line, so that the range of the vehicle is limited to the immediate vicinity of the line. There are different implementations for line navigation:

- *Electromagnetic Guidance* or *Electromagnetic Leader Cable.*

- *Reflecting Tape Guidance* (also called *Optical Tape Guidance*).

- *Ferrite Painted Guidance*, which uses ferrite magnet powder painted on the floor [Tsumura, 1986].

These techniques have been in use for many years in industrial automation tasks. Vehicles using these techniques are generally called *Automatic Guided Vehicles* (AGVs).

In this book we don't address these methods in detail, because they do not allow the vehicle to move freely — the main feature that sets mobile robots apart from AGVs. However, two recently introduced variations of the line navigation approach are of interest for mobile robots. Both techniques are based on the use of *short-lived navigational markers* (SLNM). The short-lived nature of the markers has the advantage that it is not necessary to remove the markers after use.

One typical group of applications suitable for SLNM are floor coverage applications. Examples are floor cleaning, lawn mowing, or floor surveillance. In such applications it is important for the robot to travel along adjacent paths on the floor, with minimal overlap and without "blank" spots. With the methods discussed here, the robot could conceivably mark the outside border of the path, and trace that border line in a subsequent run. One major limitation of the current state-of-the-art is that they permit only very slow travel speeds: on the order of under 10 mm/s (0.4 in/s).

7.4.1 Thermal Navigational Marker

Kleeman [1992], Kleeman and Russell [1993], and Russell [1993] report on a pyroelectric sensor that has been developed to detect thermal paths created by heating the floor with a quartz halogen bulb. The path is detected by a pyroelectric sensor based on lithium-tantalate. In order to generate a differential signal required for path following, the position of a single pyroelectric sensor is toggled between two sensing locations 5 centimeters (2 in) apart. An aluminum enclosure screens the sensor from ambient infrared light and electromagnetic disturbances. The 70 W quartz halogen bulb used in this system is located 30 millimeters (1-3/16 in) above the floor.

The volatile nature of this path is both advantageous and disadvantageous: since the heat trail disappears after a few minutes, it also becomes more difficult to detect over time. Kleeman and Russell approximated the temperature distribution T at a distance d from the trail and at a time t after laying the trail as

$$T(d,t) = A(t) \, e^{-(d/w)^2} \tag{7.2}$$

where $A(t)$ is a time-variant intensity function of the thermal path.

In a controlled experiment two robots were used. One robot laid the thermal path at a speed of 10 mm/s (0.4 in/s), and the other robot followed that path at about the same speed. Using a control scheme based on a Kalman filter, thermal paths could be tracked up to 10 minutes after being laid on a vinyl tiled floor. Kleeman and Russell remarked that the thermal footprint of peoples' feet could contaminate the trail and cause the robot to lose track.

7.4.2 Volatile Chemicals Navigational Marker

This interesting technique is based on laying down an odor trail and using an *olfactory*[1] sensor to allow a mobile robot to follow the trail at a later time. The technique was described by Deveza et

---

[1] relating to, or contributing to the sense of smell (The American Heritage Dictionary of the English Language, Third Edition is licensed from Houghton Mifflin Company. Copyright © 1992 by Houghton Mifflin Company. All rights reserved).

al. [1993] and Russell et al. [1994], and the experimental system was further enhanced as described by Russell [1995a; 1995b] at Monash University in Australia. Russell's improved system comprises a custom-built robot (see Figure 7.10) equipped with an odor-sensing system. The sensor system uses controlled flows of air to draw odor-laden air over a sensor crystal. The quartz crystal is used as a sensitive balance to weigh odor molecules. The quartz crystal has a coating with a specific affinity for the target odorant; molecules of that odorant attach easily to the coating and thereby increase the total mass of the crystal. While the change of mass is extremely small, it suffices to change the resonant frequency of the crystal. A 68HC11 microprocessor is used to count the crystal's frequency, which is in the kHz region. A change of frequency is indicative of odor concentration. In Rus-

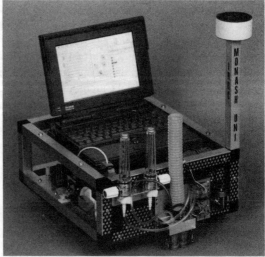

Figure 7.10: The odor-laying/odor-sensing mobile robot was developed at Monash University in Australia. The olfactory sensor is seen in front of the robot. At the top of the vertical boom is a magnetic compass. (Courtesy of Monash University).

sell's system two such sensors are mounted at a distance of 30 millimeters (1-3/16 in) from each other, to provide a differential signal that can then be used for path tracking.

For laying the odor trail, Russell used a modified felt-tip pen. The odor-laden agent is camphor, dissolved in alcohol. When applied to the floor, the alcohol evaporates quickly and leaves a 10 millimeter (0.4 in) wide camphor trail. Russell measured the response time of the olfactory sensor by letting the robot cross an odor trail at angles of 90 and 20 degrees. The results of that test are shown in Figure 7.11. Currently, the foremost limitation of Russell's volatile chemical navigational marker is the robot's slow speed of 6 mm/s (1/4 in/s).

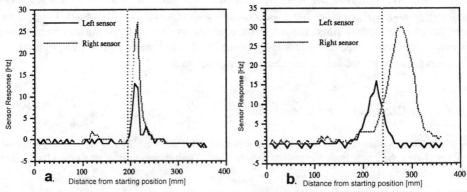

Figure 7.11: Odor sensor response as the robot crosses a line of camphor set at an angle of a. 90° and b. 20° to the robot path. The robots speed was 6 mm/s (1/4 in/s) in both tests. (Adapted with permission from Russell [1995].)

7.5 Summary

Artificial landmark detection methods are well developed and reliable. By contrast, *natural* landmark navigation is not sufficiently developed yet for reliable performance under a variety of conditions. A survey of the market of commercially available *natural* landmark systems produces only a few. One is TRC's vision system that allows the robot to localize itself using rectangular and circular ceiling lights [King and Weiman, 1990]. Cyberworks has a similar system [Cyberworks]. It is generally very difficult to develop a feature-based landmark positioning system capable of detecting different natural landmarks in different environments. It is also very difficult to develop a system that is capable of using many different types of landmarks.

We summarize the characteristics of landmark-based navigation as follows:

- Natural landmarks offer flexibility and require no modifications to the environment.

- Artificial landmarks are inexpensive and can have additional information encoded as patterns or shapes.

- The maximal distance between robot and landmark is substantially shorter than in active beacon systems.

- The positioning accuracy depends on the distance and angle between the robot and the landmark. Landmark navigation is rather inaccurate when the robot is further away from the landmark. A higher degree of accuracy is obtained only when the robot is near a landmark.

- Substantially more processing is necessary than with active beacon systems.

- Ambient conditions, such as lighting, can be problematic; in marginal visibility, landmarks may not be recognized at all or other objects in the environment with similar features can be mistaken for a legitimate landmark.

- Landmarks must be available in the work environment around the robot.

- Landmark-based navigation requires an approximate starting location so that the robot knows where to look for landmarks. If the starting position is not known, the robot has to conduct a time-consuming search process.

- A database of landmarks and their location in the environment must be maintained.

- There is only limited commercial support for this type of technique.

CHAPTER 8
MAP-BASED POSITIONING

Map-based positioning, also known as "map matching," is a technique in which the robot uses its sensors to create a map of its local environment. This local map is then compared to a global map previously stored in memory. If a match is found, then the robot can compute its actual position and orientation in the environment. The prestored map can be a CAD model of the environment, or it can be constructed from prior sensor data.

The basic procedure for map-based positioning is shown in Figure 8.1.

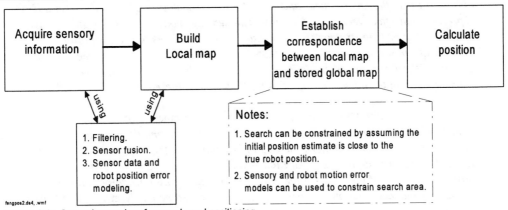

Figure 8.1: General procedure for map-based positioning.

The main advantages of map-based positioning are as follows.

- This method uses the naturally occurring structure of typical indoor environments to derive position information without modifying the environment.

- Map-based positioning can be used to generate an updated map of the environment. Environment maps are important for other mobile robot tasks, such as global path planning or the avoidance of "local minima traps" in some local obstacle avoidance methods.

- Map-based positioning allows a robot to learn a new environment and to improve positioning accuracy through exploration.

Disadvantages of map-based positioning are the specific requirements for satisfactory navigation. For example, map-based positioning requires that:

- there be enough stationary, easily distinguishable features that can be used for matching,

- the sensor map be accurate enough (depending on the tasks) to be useful,

- a significant amount of sensing and processing power be available.

One should note that currently most work in map-based positioning is limited to laboratory settings and to relatively simple environments.

8.1   Map Building

There are two fundamentally different starting points for the map-based positioning process. Either there is a pre-existing map, or the robot has to build its own environment map. Rencken [1993] defined the map building problem as the following: *"Given the robot's position and a set of measurements, what are the sensors seeing?"* Obviously, the map-building ability of a robot is closely related to its sensing capacity.

Talluri and Aggarwal [1993] explained:

"The position estimation strategies that use map-based positioning rely on the robot's ability to sense the environment and to build a representation of it, and to use this representation effectively and efficiently. The sensing modalities used significantly affect the map making strategy. Error and uncertainty analyses play an important role in accurate position estimation and map building. It is important to take explicit account of the uncertainties; modeling the errors by probability distributions and using Kalman filtering techniques are good ways to deal with these errors explicitly."

Talluri and Aggarwal [1993] also summarized the basic requirements for a map:

"The type of spatial representation system used by a robot should provide a way to incorporate consistently the newly sensed information into the existing world model. It should also provide the necessary information and procedures for estimating the position and pose of the robot in the environment. Information to do path planning, obstacle avoidance, and other navigation tasks must also be easily extractable from the built world model."

Hoppen et al. [1990] listed the three main steps of sensor data processing for map building:

1. Feature extraction from raw sensor data.

2. Fusion of data from various sensor types.

3. Automatic generation of an environment model with different degrees of abstraction.

And Crowley [1989] summarized the construction and maintenance of a composite local world model as a three-step process:

1. Building an abstract description of the most recent sensor data (a sensor model).

2. Matching and determining the correspondence between the most recent sensor models and the current contents of the composite local model.

3. Modifying the components of the composite local model and reinforcing or decaying the confidences to reflect the results of matching.

A problem related to map-building is "autonomous exploration." In order to build a map, the robot must explore its environment to map uncharted areas. Typically it is assumed that the robot begins its exploration without having any knowledge of the environment. Then, a certain motion strategy is followed which aims at maximizing the amount of charted area in the least amount of time. Such a motion strategy is called exploration strategy, and it depends strongly on the kind of sensors used. One example for a simple exploration strategy based on a lidar sensor is given by [Edlinger and Puttkamer, 1994].

8.1.1 Map-Building and Sensor Fusion

Many researchers believe that no single sensor modality alone can adequately capture all relevant features of a real environment. To overcome this problem, it is necessary to combine data from different sensor modalities, a process known as *sensor fusion*. Here are a few examples:

- Buchberger et al. [1993] and Jörg [1994; 1995] developed a mechanism that utilizes heterogeneous information obtained from a laser-radar and a sonar system in order to construct a reliable and complete world model.

- Courtney and Jain [1994] integrated three common sensing sources (sonar, vision, and infrared) for sensor-based spatial representation. They implemented a feature-level approach to sensor fusion from multisensory grid maps using a mathematical method based on *spatial moments* and *moment invariants*, which are defined as follows:

The two-dimensional $(p+q)$th order spacial moments of a grid map $G(x,y)$ are defined as

$$m_{pq} = \sum_x \sum_y x^p y^q G(x,y) \qquad p,q = 0,1,2,...$$ (8.1)

Using the centroid, translation-invariant central moments (moments don't change with the translation of the grid map in the world coordinate system) are formulated:

$$\mu_{pq} = \sum_x \sum_y (x-\bar{x})^p (y-\bar{y})^q G(x,y)$$ (8.2)

From the second- and third-order central moments, a set of seven moment invariants that are independent of translation, rotation, and scale can be derived. A more detailed treatment of spatial moments and moment invariants is given in [Gonzalez and Wintz, 1977].

8.1.2 Phenomenological vs. Geometric Representation, Engelson and McDermott [1992]

Most research in sensor-based map building attempts to minimize mapping errors at the earliest stage — when the sensor data is entered into the map. Engelson and McDermott [1992] suggest that this methodology will reach a point of diminishing returns, and hence further research should focus on explicit error detection and correction. The authors observed that the geometric approach attempts to build a more-or-less detailed geometric description of the environment from perceptual data. This has the intuitive advantage of having a reasonably well-defined relation to the real world. However, there is, as yet, no truly satisfactory representation of uncertain geometry, and it is unclear whether the volumes of information that one could potentially gather about the shape of the world are really useful.

To overcome this problem Engelson and McDermott suggested the use of a *topological* approach that constitutes a *phenomenological representation* of the robot's potential interactions with the world, and so directly supports navigation planning. Positions are represented relative to local reference frames to avoid unnecessary accumulation of relative errors. Geometric relations between frames are also explicitly represented. New reference frames are created whenever the robot's position uncertainty grows too high; frames are merged when the uncertainty between them falls sufficiently low. This policy ensures locally bounded uncertainty. Engelson and McDermott showed that such error correction can be done without keeping track of all mapping decisions ever made. The methodology makes use of the environmental structure to determine the essential information needed to correct mapping errors. The authors also showed that it is not necessary for the decision that caused an error to be specifically identified for the error to be corrected. It is enough that the *type* of error can be identified. The approach has been implemented only in a simulated environment, where the effectiveness of the phenomenological representation was demonstrated.

8.2   Map Matching

One of the most important and challenging aspects of map-based navigation is *map matching*, i.e., establishing the correspondence between a current *local map* and the stored global map [Kak et al., 1990]. Work on map matching in the computer vision community is often focused on the general problem of matching an image of arbitrary position and orientation relative to a model (e.g., [Talluri and Aggarwal, 1993]). In general, matching is achieved by first extracting features, followed by determination of the correct correspondence between image and model features, usually by some form of constrained search [Cox, 1991].

Such matching algorithms can be classified as either *icon-based* or *feature-based*. Schaffer et al. [1992] summarized these two approaches:

> "*Iconic-based pose estimation pairs sensory data points with features from the map, based on minimum distance. The robot pose is solved for that minimizes the distance error between the range points and their corresponding map features. The robot pose is solved [such as to] minimize the distance error between the range points and their corresponding map features. Based on the new pose, the correspondences are recomputed and the process repeats until the change in aggregate distance error between points and line segments falls below a threshold. This algorithm differs from the feature-based method in that it matches every range data point to the map rather than corresponding the range data into a small set of features to be matched to the map. The feature-based estimator, in general, is faster than the iconic estimator and does not require a good initial heading estimate. The iconic estimator can use fewer points than the feature-based estimator, can handle less-than-ideal environments, and is more accurate. Both estimators are robust to some error in the map.*"

Kak et al. [1990] pointed out that one problem in map matching is that the sensor readings and the world model may be of different formats. One typical solution to this problem is that the approximate position based on odometry is utilized to generate (from the prestored global model), an estimated visual scene that would be "seen" by robot. This estimated scene is then matched against the actual scene viewed by the onboard sensors. Once the matches are established between the features of the two images (expected and actual), the position of the robot can be estimated with reduced uncertainty. This approach is also supported by Rencken [1994], as will be discussed in more detail below.

In order to match the current sensory data to the stored environment model reliably, several features must be used simultaneously. This is particularly true for a range image-based system since the types of features are limited to a range image map. Long walls and edges are the most commonly used features in a range image-based system. In general, the more features used in one match, the less likely a mismatch will occur, but the longer it takes to process. A realistic model for the odometry and its associated uncertainty is the basis for the proper functioning of a map-based positioning system. This is because the feature detection as well as the updated position calculation rely on odometric estimates [Chenavier and Crowley, 1992].

8.2.1 Schiele and Crowley [1994]

Schiele and Crowley [1994] discussed different matching techniques for matching two occupancy grids. The first grid is the local grid that is centered on the robot and models its vicinity using the most recent sensor readings. The second grid is a global model of the environment furnished either by learning or by some form of computer-aided design tool. Schiele and Crowley propose that two representations be used in environment modeling with sonars: *parametric primitives* and an *occupancy grid*. Parametric primitives describe the limits of free space in terms of segments or surfaces defined by a list of parameters. However, noise in the sensor signals can make the process of grouping sensor readings to form geometric primitives unreliable. In particular, small obstacles such as table legs are practically impossible to distinguish from noise.

Schiele and Crowley discuss four different matches:

- Matching segment to segment as realized by comparing segments in (1) similarity in orientation, (2) collinearity, and (3) overlap.

- Matching segment to grid.

- Matching grid to segment.

- Matching grid to grid as realized by generating a mask of the local grid. This mask is then transformed into the global grid and correlated with the global grid cells lying under this mask. The value of that correlation increases when the cells are of the same state and decreases when the two cells have different states. Finally finding the transformation that generates the largest correlation value.

Schiele and Crowley pointed out the importance of designing the updating process to take into account the uncertainty of the local grid position. The correction of the estimated position of the robot is very important for the updating process particularly during exploration of unknown environments.

Figure 8.2 shows an example of one of the experiments with the robot in a hallway. Experimental results obtained by Schiele and Crowley show that the most stable position estimation results are obtained by matching segments to segments or grids to grids.

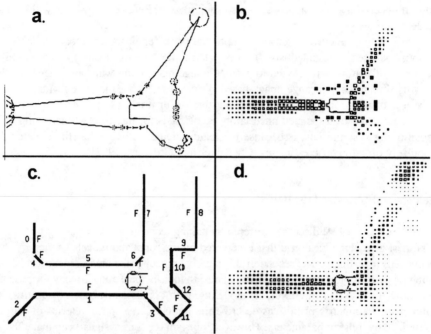

Figure 8.2: Schiele and Crowley's robot models its position in a hallway.
a. Raw ultrasonic range data projected onto external coordinates around the robot.
b. Local grid and the edge segments extracted from this grid.
c. The robot with its uncertainty in estimated position within the global grid.
d. The local grid imposed on the global grid at the position and orientation of best correspondence.
(Reproduced and adapted from [Schiele and Crowley, 1994].)

8.2.2 Hinkel and Knieriemen [1988] — The *Angle Histogram*

Hinkel and Knieriemen [1988] from the University of Kaiserslautern, Germany, developed a world-modeling method called the *Angle Histogram*. In their work they used an in-house developed lidar mounted on their mobile robot *Mobot III*. Figure 8.3 shows that lidar system mounted on *Mobot III's* successor *Mobot IV*. (Note that the photograph in Figure 8.3 is very recent; it shows *Mobot IV* on the left, and *Mobot V*, which was built in 1995, on the right. Also note that an ORS-1 lidar from ESP, discussed in Sec. 4.2.2, is mounted on *Mobot V*.)

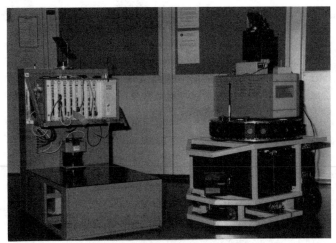

Figure 8.3: *Mobot IV* (left) and *Mobot V* (right) were both developed and built at the University of Kaiserslautern. The different *Mobot* models have served as mobile robot testbeds since the mid-eighties. (Courtesy of the University of Kaiserslautern.)

A typical scan from the in-house lidar is shown in Figure 8.4. The similarity between the scan quality of the University of Kaiserslautern lidar and that of the ORS-1 lidar (see Fig. 4.32a in Sec. 4.2.6) is striking.

The angle histogram method works as follows. First, a 360 degree scan of the room is taken with the lidar, and the resulting "hits" are recorded in a map. Then the algorithm measures the relative angle δ between any two adjacent hits (see Figure 8.5). After compensating for noise in the readings (caused by the inaccuracies in position between adjacent hits), the angle histogram shown in Figure 8.6a can be built. The uniform direction of the main walls are clearly visible as peaks in the angle histogram. Computing the histogram modulo π results in only two main peaks: one for each pair of parallel walls. This algorithm is very robust with regard to openings in the walls, such as doors and windows, or even cabinets lining the walls.

After computing the angle histogram, all angles of the hits can be normalized, resulting in the representation shown in Figure 8.6b. After this transformation, two additional histograms, one for the x- and one for the y-direction can be constructed. This time, peaks show the distance to the walls in x and y direction. During operation, new orientation and position data is updated at a rate of 4 Hz. (In conversation with Prof. Von

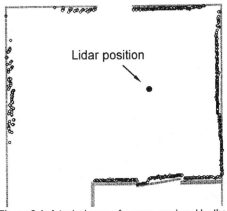

Figure 8.4: A typical scan of a room, produced by the University of Kaiserslautern's in-house developed lidar system. (Courtesy of the University of Kaiserslautern.)

Puttkamer, Director of the Mobile Robotics Laboratory at the University of Kaiserslautern, we learned that this algorithm had since been improved to yield a reliable accuracy of 0.5°.)

8.2.3 Weiß, Wetzler, and Puttkamer — More on the *Angle Histogram*

Weiß et al. [1994] conducted further experiments with the angle histogram method. Their work aimed at matching rangefinder scans from different locations. The purpose of this work was to compute the translational and rotational displacement of a mobile robot that had traveled during subsequent scans.

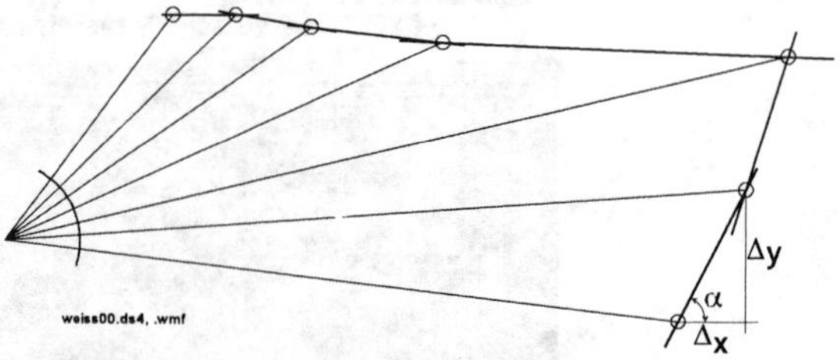

weiss00.ds4, .wmf

Figure 8.5: Calculating angles for the angle histogram. (Courtesy of [Weiß et al., 1994].)

The authors pointed out that an angle histogram is mostly invariant against rotation and translation. If only the orientation is altered between two scans, then the angle histogram of the second scan will show only a phase shift when compared to the first. However, if the position of the robot is altered, too, then the distribution of angles will also change. Nonetheless, even in that case the new angle histogram will still be a representation of the distribution of directions in the new scan. Thus, in the new angle histogram the same direction that appeared to be the local maximum in the old angle histogram will still appear as a maximum, provided the robot's displacement between the two scans was sufficiently small.

Experiments show that this approach is highly stable against noise, and even moving obstacles do not distort the result as long as they do not represent the majority of matchable data. Figure

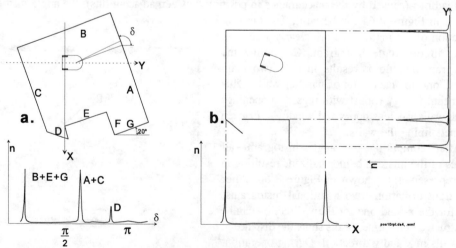

Figure 8.6:   Readings from a rotating laser scanner generate the contours of a room.
a. The angle histogram allows the robot to determine its orientation relative to the walls.
b. After normalizing the orientation of the room relative to the robot, an x-y histogram can be built form the same data points. (Adapted from [Hinkel and Knieriemen, 1988].)

8.7a shows two scans taken from two different locations. The second scan represents a rotation of +43 degrees, a translation in x-direction of +14 centimeters and a translation in y-direction of +96 centimeters. Figure 8.7b shows the angle histogram associated with the two positions. The maxima for the main directions are -24 and 19 degrees, respectively. These angles correspond to the rotation of the robot relative to the local main direction.

Definition

A cross-correlation is defined as

$$c(y) = \lim_{X \to \infty} \frac{1}{2X} \int_{-X}^{X} f(x)\,g(x+y)\,dx \ . \tag{8.3}$$

$c(y)$ is a measure of the cross-correlation between two stochastic functions regarding the phase shift y. The cross-correlation $c(y)$ will have an absolute maximum at s, if $f(x)$ is equal to $g(x+s)$. (Courtesy of [Weiß et al., 1994].)

One can thus conclude that the rotational displacement of the robot was $19° - (-24°) = +43°$. Furthermore, rotation of the first and second range plot by -24 and 19 degrees, respectively, provides the normalized x- and y-plots shown in Fig 8.7c. The cross correlation of the x translation is shown in Figure 8.7d. The maximum occurs at -35 centimeters, which corresponds to -14 centimeters in the rotated scan (Fig. 8.7a). Similarly, the y-translation can be found to be +98 centimeters in the rotated scan. Figure 8.5e shows the result of scan matching after making all rotational and translational corrections.

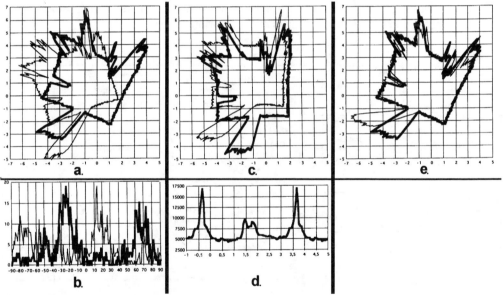

Figure 8.7: Various stages during the matching of two angle histograms. The two histograms were built from scan data taken from two different locations. (Courtesy of [Weiß et al., 1994].)
a. Two scans with rotation of +43°, x-transition of +14 cm, y-transition of +96 cm.
b. Angle histogram of the two positions.
c. Scans rotated according to the maximum of their angle histogram (+24°, -19°).
d. Cross-correlation of the x-translation (maximum at -35 cm, corresponding to -14 cm in the rotated scan).
e. x-translation correction of +14 cm; y-translation correction of -98 cm.

8.2.4 Siemens' *Roamer*

Rencken [1993; 1994] at the Siemens Corporate Research and Development Center in Munich, Germany, has made substantial contributions toward solving the boot strap problem resulting from the uncertainty in position and environment. This problem exists when a robot must move around in an unknown environment, with uncertainty in the robot's odometry-derived position. For example, when building a map of the environment, all measurements are necessarily relative to the carrier of the sensors (i.e., the mobile robot). Yet, the position of the robot itself is not known exactly, because of the errors accumulating in odometry.

Rencken addresses the problem as follows: in order to represent features "seen" by its 24 ultrasonic sensors, the robot constructs hypotheses about these features. To account for the typically unreliable information from ultrasonic sensors, features can be classified as *hypothetical*, *tentative*, or *confirmed*. Once a feature is *confirmed*, it is used for constructing the map as shown in Figure 8.8. Before the map can be updated, though, every new data point must be associated with either a plane, a corner, or an edge (and some variations of these features). Rencken devices a "hypothesis tree" which is a data structure that allows tracking of different hypotheses until a sufficient amount of data has been accumulated to make a final decision.

One further important aspect in making this decision is *feature visibility*. Based on internal models for different features, the robot's decisions are aided by a routine check on visibility. For example, the visibility of edges is smaller than that of corners. The visibility check further reduces the uncertainty and improves the robustness of the algorithm.

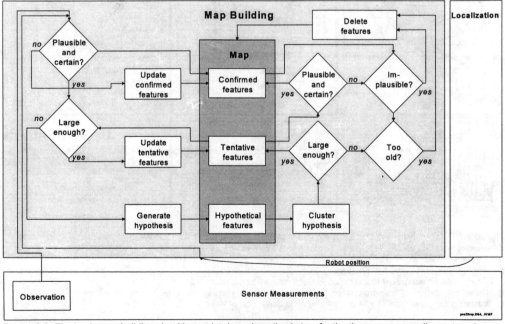

Figure 8.8: The basic map-building algorithm maintains a *hypothesis tree* for the three sensor reading categories: *hypothetical*, *tentative*, and *confirmed*. (Adapted from [Rencken, 1994].)

Based on the above methods, Rencken [1993] summarizes his method with the following procedure:
1. Predict the robot's position using odometry.
2. Predict the associated covariance of this position estimate.
3. Among the set of given features, test which feature is visible to which sensor and predict the measurement.
4. Compare the predicted measurements to the actual measurements.
5. Use the error between the validated and predicted measurements to estimate the robot's position.
6. The associated covariance of the new position estimate is also determined.

The algorithm was implemented on Siemens' experimental robot *Roamer*. In an endurance experiment, *Roamer* traveled through a highly cluttered office environment for approximately 20 minutes. During this time, the robot updated its internal position only by means of odometry and its map-building capabilities. At a relatively slow travel speed of 12 cm/s (4¾ in/s) *Roamer*'s position accuracy was periodically recorded, as shown in Table 8.1.

Table 8.1: Position and orientation errors of Siemens' *Roamer* robot in an map-building "endurance test." (Adapted from [Rencken, 1994].)

Time [min:sec]	Pos. Error [cm] (in)	Orientation error [°]
5:28	5.8 (2-1/4)	-7.5
11:57	5.3 (2)	-6.2
14:53	5.8 (2-1/4)	0.1
18:06	4.0 (1-1/2)	-2.7
20:12	2.5 (1)	3.0

8.3 Geometric and Topological Maps

In map-based positioning there are two common representations: *geometric* and *topological* maps. A geometric map represents objects according to their absolute geometric relationships. It can be a grid map, or a more abstracted map, such as a line map or a polygon map. In map-based positioning, sensor-derived geometric maps must be matched against a global map of a large area. This is often a formidable difficulty because of the robot's position error. By contrast, the *topological* approach is based on recording the geometric relationships between the observed features rather than their absolute position with respect to an arbitrary coordinate frame of reference. The resulting presentation takes the form of a graph where the nodes represent the observed features and the edges represent the relationships between the features. Unlike geometric maps, topological maps can be built and maintained without any estimates for the position of the robot. This means that the errors in this representation will be independent of any errors in the estimates for the robot position [Taylor, 1991]. This approach allows one to integrate large area maps without suffering from the accumulated odometry position error since all connections between nodes are relative, rather than absolute. After the map has been established, the positioning process is essentially the process of matching a local map to the appropriate location on the stored map.

8.3.1 Geometric Maps for Navigation

There are different ways for representing geometric map data. Perhaps the simplest way is an occupancy grid-based map. The first such map (in conjunction with mobile robots) was the *Certainty Grid* developed by Moravec and Elfes, [1985]. In the Certainty Grid approach, sensor readings are placed into the grid by using probability profiles that describe the algorithm's certainty about the existence of objects at individual grid cells. Based on the Certainty Grid approach, Borenstein and Koren [1991] refined the method with the Histogram Grid, which derives a pseudo-probability distribution out of the motion of the robot [Borenstein and Koren, 1991]. The Histogram Grid method is now widely used in many mobile robots (see for example [Buchberger et al., 1993; Congdon et al., 1993; Courtney and Jain, 1994; Stuck et al., 1994; Wienkop et al., 1994].)

A measure of the goodness of the match between two maps and a trial displacement and rotation is found by computing the sum of products of corresponding cells in the two maps [Elfes, 1987]. Range measurements from multiple points of view are symmetrically integrated into the map. Overlapping empty volumes reinforce each other and serve to condense the range of the occupied volumes. The map definition improves as more readings are added. The method deals effectively with clutter and can be used for motion planning and extended landmark recognition.

The advantages of occupancy grid-based maps are that they:
- allow higher density than stereo maps,
- require less computation and can be built more quickly,
- allow for easy integration of data from different sensors, and
- can be used to express statistically the confidence in the correctness of the data [Raschke and Borenstein, 1990].

The disadvantages of occupancy grid-based maps are that they:
- have large uncertainty areas associated with the features detected,
- have difficulties associated with active sensing [Talluri and Aggarwal, 1993],
- have difficulties associated with modeling of dynamic obstacles, and
- require a more complex estimation process for the robot vehicle [Schiele and Crowley, 1994].

In the following sections we discuss some specific examples for occupancy grid-based map matching.

8.3.1.1 Cox [1991]

One typical grid-map system was implemented on the mobile robot *Blanche* [Cox, 1991]. This positioning system is based on matching a local grid map to a global line segment map. *Blanche* is designed to operate autonomously within a structured office or factory environment without active or passive beacons. *Blanche's* positioning system consists of :

- an *a priori* map of its environment, represented as a collection of discrete line segments in the plane,
- a combination of odometry and a rotating optical range sensor to sense the environment,
- an algorithm for matching the sensory data to the map, where matching is constrained by assuming that the robot position is roughly known from odometry, and
- an algorithm to estimate the precision of the corresponding match/correction that allows the correction to be combined optimally (in a maximum likelihood sense) with the current odometric position to provide an improved estimate of the vehicle's position.

The operation of Cox's map-matching algorithm (item 2, above) is quite simple. Assuming that the sensor hits are near the actual objects (or rather, the lines that represent the objects), the distance between a hit and the *closest* line is computed. This is done for each point, according to the procedure in Table 8.2 (from [Cox, 1991]).

Table 8.2: Procedure for implementing Cox's [1991] map-matching algorithm.

1. For each point in the image, find the line segment in the model that is nearest to the point. Call this the *target*.
2. Find the congruence that minimizes the total squared distance between the image points and their target lines.
3. Move the points by the congruence found in step 2.
4. Repeat steps 1 to 3 until the procedure converges.

Figure 8.9 shows how the algorithm works on a set of real data. Figure 8.9a shows the line model of the contours of the office environment (solid lines). The dots show hits by the range sensor. This scan was taken while the robot's position estimate was offset from its true position by 2.75 meters (9 ft) in the x-direction and 2.44 meters (8 ft) in the y-direction. A very small orientation error was also present. After running the map-matching procedure in Table 8.2, the robot corrected its internal position, resulting in the very good match between sensor data and line model, shown in Figure 8.9b.

In a longer run through corridors and junctions *Blanche* traveled at various slow speeds, on the order of 5 cm/s (2 in/s). The maximal deviation of its computed position from the actual position was said to be 15 centimeters (6 in).

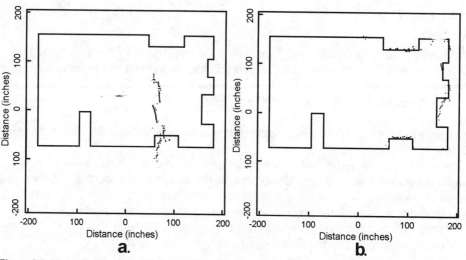

Figure 8.9: Map and range data a. before registration and b. after registration. (Reproduced and adapted from [Cox, 1991], © 1991 IEEE.)

Discussion

With the grid-map system used in *Blanche,* generality has been sacrificed for robustness and speed. The algorithm is intrinsically robust against incompleteness of the image. Incompleteness of the model is dealt with by deleting any points whose distance to their target segments exceed a certain limit. In Cox's approach, a reasonable heuristic used for determining correspondence is the minimum Euclidean distance between the model and sensed data. Gonzalez et al. [1992] comment that this assumption is valid only as long as the displacement between the sensed data and the model is sufficiently small. However, this minimization problem is inherently non-linear but is linearized assuming that the rotation angle is small. To compensate for the error introduced due to linearization, the computed position correction is applied to the data points, and the process is repeated until no significant improvement can be obtained [Jenkin et al., 1993].

8.3.1.2 Crowley [1989]

Crowley's [1989] system is based on matching a local line segment map to a global line segment map. Crowley develops a model for the uncertainty inherent in ultrasonic range sensors, and he describes a method for the projection of range measurements onto external Cartesian coordinates. Crowley develops a process for extracting line segments from adjacent collinear range measurements, and he presents a fast algorithm for matching these line segments to a model of the *geometric limits for the free-space* of the robot. A side effect of matching sensor-based observations to the model is a correction to the estimated position of the robot at the time that the observation was made. The projection of a segment into the external coordinate system is based on the estimate of the position of the vehicle. Any uncertainty in the vehicle's estimated position must be included in the uncertainty of the segment before matching can proceed. This uncertainty affects both the position and orientation of the line segment. As each segment is obtained from the sonar data, it is matched to the composite model. Matching is a process of comparing each of the segments in the composite local model against the observed segment, to allow detection of similarity in orientation, collinearity, and overlap. Each of these tests is made by comparing one of the parameters in the segment representation:

a. **Orientation** The square of the difference in orientation of the two candidates must be smaller than the sum of the variances.

b. **Alignment** The square of the difference of the distance from the origin to the two candidates must be smaller than the sum of the corresponding variance.

c. **Overlap** The difference of the distance between centerpoints to the sum of the half lengths must be smaller than a threshold.

The longest segment in the composite local model which passes all three tests is selected as the matching segment. The segment is then used to correct the estimated position of the robot and to update the model. An explicit model of uncertainty using covariance and Kalman filtering provides a tool for integrating noisy and imprecise sensor observations into the model of the geometric limits for the free space of a vehicle. Such a model provides a means for a vehicle to maintain an estimate of its position as it travels, even in the case where the environment is unknown.

Figure 8.10 shows the model of the ultrasonic range sensor and its uncertainties (shown as the hatched area A). The length of A is given by the uncertainties in robot orientation σ_w and the width is given by the uncertainty in depth σ_D. This area is approximated by an ellipse with the major and minor axis given by σ_w and σ_D.

Figure 8.10: Model of the ultrasonic range sensor and its uncertainties. (Adapted from [Crowley, 1989].)

Figure 8.11 shows a vehicle with a circular uncertainty in position of 40 centimeters (16 in) detecting a line segment. The ultrasonic readings are illustrated as circles with a radius determined by its uncertainty as defined in Figure 8.10. The detected line segment is illustrated by a pair of parallel lines. (The actual line segment can fall anywhere between the two lines. Only uncertainties associated with sonar readings are considered here.)

Figure 8.11b shows the segment after the uncertainty in the robot's position has been added to the segment uncertainties. Figure 8.11c shows the uncertainty in position after correction by matching a model segment. The position uncertainty of the vehicle is reduced to an ellipse with a minor axis of approximately 8 centimeters (3.15 in).

In another experiment, the robot was placed inside the simple environment shown in Figure 8.12. Segment 0 corresponds to a wall covered with a textured wall-paper. Segment 1

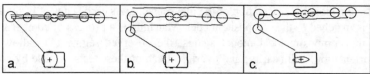

Figure 8.11:
a. A vehicle with a position uncertainty of 40 cm (15.7 in), as shown by the circle
    around the centerpoint (cross), is detecting a line segment.
b. The boundaries for the line segment grow after adding the uncertainty for the
    robot's position.
c. After correction by matching the segment boundaries with a stored map segment,
    the uncertainty of the robot's position is reduced to about 8 cm (3.15 in) as shown
    by the squat ellipse around the robot's center (cross).
Courtesy of [Crowley, 1989].

corresponds to a metal cabinet with a sliding plastic door. Segment 2 corresponds to a set of chairs
pushed up against two tables. The robot system has no *a priori* knowledge of its environment. The
location and orientation at which the system was started were taken as the origin and x-axis of the
world coordinate system. After the robot has run three cycles of ultrasonic acquisition, both the
estimated position and orientation of the vehicle were set to false values. Instead of the correct
position ($x = 0$, $y = 0$, and $\theta = 0$), the position was set to $x = 0.10$ m, $y = 0.10$ m, and the
orientation was set to 5 degrees. The uncertainty was set to a standard deviation of 0.2 meters in
x and y, with a uncertainty in orientation of 10 degrees. The system was then allowed to detect the
"wall" segments around it. The resulting estimated position and covariance is listed in Table 8.3.

Table 8.3: Experimental results with Crowley's map-matching method. Although initially placed in an incorrect position, the robot corrects its position error with every additional wall segment scanned.

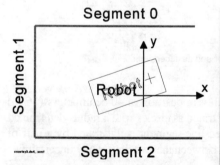

Figure 8.12: Experimental setup for testing Crowley's map-matching method. Initially, the robot is intentionally set-off from the correct starting position.

Initial estimated position (with deliberate initial error)	x,y,θ = (0.100, 0.100, 5.0)		
Covariance	0.040	0.000	0.000
	0.000	0.040	0.000
	0.000	0.000	100.0
After match with segment 0 estimated position:	x,y,θ = (0.102, 0.019, 1.3)		
Covariance	0.039	0.000	0.000
	0.000	0.010	0.000
	0.000	0.000	26.28
After match with segment 1 estimated position:	x,y,θ = (0.033, 0.017, 0.20)		
Covariance	0.010	0.000	0.000
	0.000	0.010	0.000
	0.000	0.000	17.10

8.3.1.3 Adams and von Flüe

The work by Adams and von Flüe follows the work by Leonard and Durrant-Whyte [1990] in using an approach to mobile robot navigation that unifies the problems of obstacle detection, position estimation, and map building in a common multi-target tracking framework. In this approach a mobile robot continuously tracks naturally occurring indoor targets that are subsequently treated as "beacons." Predicted targets (i.e., those found from the known environmental map) are tracked in order to update the position of the vehicle. Newly observed targets (i.e., those that were *not* predicted) are caused by unknown environmental features or obstacles from which new tracks are initiated, classified, and eventually integrated into the map.

Adams and von Flüe implemented the above technique using real sonar data. The authors note that a good sensor model is crucial for this work. For this reason, and in order to predict the expected observations from the sonar data, they use the sonar model presented by Kuc and Siegel [1987].

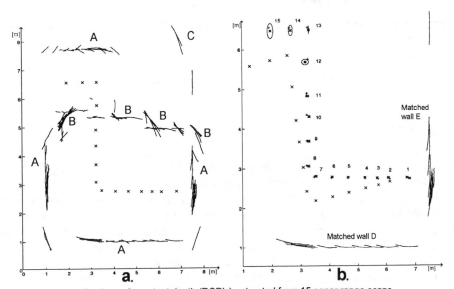

Figure 8.13: a. Regions of constant depth (RCD's) extracted from 15 sonar range scans.
b. True (x), odometric (+), and estimated (*) positions of the mobile robot using two
planar (wall) "beacons" for localization. (Courtesy of Adams and von Flüe.)

Figure 8.13a shows *regions of constant depth* (RCDs) [Kuc and Siegel, 1987] that were extracted from 15 sonar scans recorded from each of the locations marked "×."

The model from Kuc and Siegel's work suggests that RCDs such as those recorded at the positions marked A in Figure 8.13a correspond to planar surfaces; RCDs marked B rotate about a point corresponding to a 90 degree corner and RCDs such as C, which cannot be matched, correspond to multiple reflections of the ultrasonic wave.

Figure 8.13b shows the same mobile robot run as Figure 8.13a, but here the robot computes its position from *two* sensed "beacons," namely the wall at D and the wall at E in the right-hand scan in Figure 8.13b. It can be seen that the algorithm is capable of producing accurate positional estimates of the robot, while simultaneously building a map of its sensed environment as the robot becomes more confident of the nature of the features.

8.3.2 Topological Maps for Navigation

Topological maps are based on recording the *geometric relationships* between the observed features rather than their absolute position with respect to an arbitrary coordinate frame of reference. Kortenkamp and Weymouth [1994] defined the two basic functions of a topological map:

a. **Place Recognition** With this function, the current location of the robot in the environment is determined. In general, a description of the place, or node in the map, is stored with the place. This description can be abstract or it can be a local sensory map. At each node, matching takes place between the sensed data and the node description.

b. **Route Selection** With this function, a path from the current location to the goal location is found.

The following are brief descriptions of specific research efforts related to topological maps.

8.3.2.1 Taylor [1991]

Taylor, working with stereo vision, observed that each local stereo map may provide good estimates for the relationships between the observed features. However, because of errors in the estimates for the robot's position, local stereo maps don't necessarily provide good estimates for the coordinates of these features with respect to the base frame of reference. The recognition problem in a topological map can be reformulated as a *graph-matching* problem where the objective is to find a set of features in the relational map such that the relationships between these features match the relationships between the features on the object being sought. Reconstructing Cartesian maps from relational maps involves minimizing a non-linear objective function with multiple local minima.

8.3.2.2 Courtney and Jain [1994]

A typical example of a topological map-based approach is given by Courtney and Jain [1994]. In this work the coarse position of the robot is determined by classifying the map description. Such classification allows the recognition of the workspace region that a given map represents. Using data collected from 10 different rooms and 10 different doorways in a building (see Fig. 8.14), Courtney and Jain estimated a 94 percent recognition rate of the rooms and a 98 percent recognition rate of the doorways. Courtney and Jain concluded that coarse position estimation, or place recognition, in indoor domains is possible through classification of grid-based maps. They developed a paradigm wherein pattern classification techniques are applied to the task of mobile robot localization. With this paradigm the robot's workspace is represented as a set of grid-based maps interconnected via topological relations. This representation scheme was chosen over a single global map in order to avoid inaccuracies due to cumulative dead-reckoning error. Each region is represented by a set of multi-sensory grid maps, and feature-level sensor fusion is accomplished through extracting spatial descriptions from these maps. In the navigation phase, the robot localizes itself by comparing features extracted from its map of the

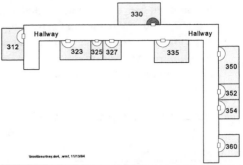

Figure 8.14: Based on datasets collected from 10 different rooms and 10 different doorways in a building, Courtney and Jain estimate a 94 percent recognition rate of the rooms and a 98 percent recognition rate of the doorways. (Adapted from [Courtney and Jain, 1994].)

current locale with representative features of known locales in the environment. The goal is to recognize the current locale and thus determine the workspace region in which the robot is present.

8.3.2.3 Kortenkamp and Weymouth [1993]

Kortenkamp and Weymouth implemented a *cognitive* map that is based on a topological map. In their topological map, instead of looking for places that are locally distinguishable from other places and then storing the distinguishing features of the place in the route map, their algorithm looks for places that mark the transition between one space in the environment and another space (gateways). In this algorithm sonar and vision sensing is combined to perform place recognition for better accuracy in recognition, greater resilience to sensor errors, and the ability to resolve

ambiguous places. Experimental results show excellent recognition rate in a well-structured environment. In a test of seven gateways, using *either* sonar or vision only, the system correctly recognized only four out of seven places. However, when sonar and vision were combined, all seven places were correctly recognized. Figure 8.15 shows the experimental space for place recognition. Key locations are marked in capital letters. Table 8.4a and Table 8.4b show the probability for each place using only vision and sonar, respectively. Table 8.4c shows the combined probabilities (vision and sonar) for each place.   In spite of the good results evident from Table 8.4c, Kortenkamp and Weymouth pointed out several drawbacks of their system:

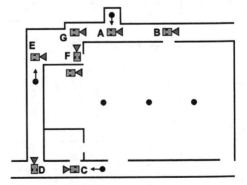

Figure 8.15: An experiment to determine if the robot can detect the same place upon return at a later time. In this case, multiple paths through the place can be "linked" together to form a network. (Adapted from [Kortenkamp and Weymouth, 1994].)

- The robot requires several initial, guided traversals of a route in order to acquire a stable set of location cues so that it can navigate autonomously.
- Acquiring, storing, and matching visual scenes is very expensive, both in computation time and storage space.
- The algorithm is restricted to highly structured, orthogonal environments.

Table 8.4a: Probabilities for each place using only vision.

	Stored Places						
	A	B	C	D	E	F	G
A	0.43	0.09	0.22	0.05	0.05	0.1	0.06
B	0.05	0.52	0.21	0.06	0.05	0.05	0.05
C	0.10	0.12	0.36	0.2	0.04	0.13	0.04
D	0.14	0.05	0.24	0.43	0.05	0.04	0.05
E	0.14	0.14	0.14	0.14	0.14	0.14	0.14
F	0.14	0.14	0.14	0.16	0.14	0.14	0.14
G	0.14	0.14	0.14	0.14	0.14	0.14	0.14

Table 8.4b: Probabilities for each place using only sonar.

	Stored Places						
	A	B	C	D	E	F	G
A	0.82	0.04	0.04	0.04	0.04	0	0
B	0.02	0.31	0.31	0.31	0.06	0	0
C	0.02	0.31	0.31	0.31	0.06	0	0
D	0.02	0.31	0.31	0.31	0.61	0	0
E	0.04	0.12	0.12	0.12	0.61	0	0
F	0	0	0	0	0	0.90	0.10
G	0	0	0	0	0	0.10	0.90

Table 8.4c: Combined probabilities (vision and sonar) for each place.

	Stored Places						
	A	B	C	D	E	F	G
A	0.95	0.01	0.02	0.01	0.01	0	0
B	0	0.65	0.26	0.07	0.01	0	0
C	0	0.17	0.52	0.29	0.01	0	0
D	0.01	0.07	0.33	0.58	0.01	0	0
E	0.04	0.12	0.12	0.12	0.61	0	0
F	0	0	0	0	0	0.90	0.1
G	0	0	0	0	0	0.09	0.91

8.4 Summary

Map-based positioning is still in the research stage. Currently, this technique is limited to laboratory settings and good results have been obtained only in well-structured environments. It is difficult to judge how the performance of a laboratory robot scales up to a real world application. Kortenkamp and Weymouth [1994] noted that very few systems tested on real robots are tested under realistic conditions with more than a handful of places.

We summarize relevant characteristics of map-based navigation systems as follows:

Map-based navigation systems:
- are still in the research stage and are limited to laboratory settings,
- have not been tested extensively in real-world environments,
- require a significant amount of processing and sensing capability,
- need extensive processing, depending on the algorithms and resolution used,
- require initial position estimates from odometry in order to limit the initial search for features to a smaller area.

There are several critical issues that need to be developed further:
- Sensor selection and sensor fusion for specific applications and environments.
- Accurate and reliable algorithms for matching local maps to the stored map.
- Good error models of sensors and robot motion.
- Good algorithms for integrating local maps into a global map.

APPENDIX A
A WORD ON KALMAN FILTERS

The most widely used method for sensor fusion in mobile robot applications is the *Kalman filter*. This filter is often used to combine all measurement data (e.g., for fusing data from different sensors) to get an optimal estimate in a statistical sense. If the system can be described with a linear model and both the system error and the sensor error can be modeled as white Gaussian noise, then the Kalman filter will provide a unique statistically optimal estimate for the fused data. This means that under certain conditions the Kalman filter is able to find the best estimates based on the "correctness" of each individual measurement.

The calculation of the Kalman filter is done recursively, i.e., in each iteration, only the newest measurement and the last estimate will be used in the current calculation, so there is no need to store all the previous measurements and estimates. This characteristic of the Kalman filter makes it appropriate for use in systems that don't have large data storage capabilities and computing power. The measurements from a group of *n* sensors can be fused using a Kalman filter to provide both an estimate of the current state of a system and a prediction of the future state of the system.

The inputs to a Kalman filter are the system measurements. The *a priori* information required are the system dynamics and the noise properties of the system and the sensors. The output of the Kalman filter is the estimated system state and the *innovation* (i.e., the difference between the predicted and observed measurement). The innovation is also a measure for the performance of the Kalman filter.

At each step, the Kalman filter generates a state estimate by computing a weighted average of the predicted state (obtained from the system model) and the innovation. The weight used in the weighted average is determined by the covariance matrix, which is a direct indication of the error in state estimation. In the simplest case, when all measurements have the same accuracy and the measurements are the states to be estimated, the estimate will reduce to a simple average, i.e., a weighted average with all weights equal. Note also that the Kalman filter can be used for systems with time-variant parameters.

The *extended Kalman filter* is used in place of the conventional Kalman filter if the system model is potentially numerically instable or if the system model is not approximately linear. The extended Kalman filter is a version of the Kalman filter that can handle non-linear dynamics or non-linear measurement equations, or both [Abidi and Gonzalez, 1992].

APPENDIX B
UNIT CONVERSIONS AND ABBREVIATIONS

To convert from	To	Multiply by
(Angles)		
degrees (°)	radian (rad)	0.01745
radian (rad)	degrees (°)	57.2958
milliradians (mrad)	degrees (°)	0.0573
(Length)		
inch (in)	meter (m)	0.0254
inch (in)	centimeter (cm)	2.54
inch (in)	millimeter (mm)	25.4
foot (ft)	meter (m)	30.48
mile (mi), (U.S. statute)	meter (m)	1,609
mile (mi), (international nautical)	meter (m)	1,852
yard (yd)	meter (m)	0.9144
(Area)		
$inch^2$ (in^2)	$meter^2$ (m^2)	6.4516×10^{-4}
$foot^2$ (ft^2)	$meter^2$ (m^2)	9.2903×10^{-2}
$yard^2$ (yd^2)	$meter^2$ (m^2)	0.83613
(Volume)		
$foot^3$ (ft^3)	$meter^3$ (m^3)	2.8317×10^{-2}
$inch^3$ (in^3)	$meter^3$ (m^3)	1.6387×10^{-5}
(Time)		
nanosecond (ns)	second (s)	10^{-9}
microsecond (μs)	second (s)	10^{-6}
millisecond (ms)	second (s)	10^{-3}
second (s)		
minute (min)	second (s)	60
hour (hr)	second (s)	3,600
(Frequency)		
Hertz (Hz)	cycle/second (s^{-1})	1
Kilohertz (KHz)	Hz	1,000
Megahertz (MHz)	Hz	10^6
Gigahertz (GHz)	Hz	10^9

To convert from	To	Multiply by
(Velocity)		
foot/minute (ft/min)	meter/second (m/s)	5.08×10^{-3}
foot/second (ft/s)	meter/second (m/s)	0.3048
knot (nautical mi/h)	meter/second (m/s)	0.5144
mile/hour (mi/h)	meter/second (m/s)	0.4470
mile/hour (mi/h)	kilometer/hour (km/h)	1.6093
(Mass, Weight)		
pound mass (lb)	kilogram (kg)	0.4535
pound mass (lb)	grams (gr)	453.59
ounce mass (oz)	grams (gr)	28.349
slug (lbf · s^2/ft)	kilogram (kg)	14.594
ton (2000 lbm)	kilogram (kg)	907.18
(Force)		
pound force (lbf)	newton (N)	4.4482
ounce force	newton (N)	0.2780
(Energy, Work)		
foot-pound force (ft · lbf)	joule (J)	1.3558
kilowatt-hour (kW · h)	joule (J)	3.60×10^6
(Acceleration)		
foot/second^2 (ft/s^2)	meter/second^2 (m/s^2)	0.3048
inch/second (in./s^2)	meter/second^2 (m/s^2)	2.54×10^{-2}
(Power)		
foot-pound/minute (ft · lbf/min)	watt (W)	2.2597×10^{-2}
horsepower (550 ft · lbf/s)	watt (W)	745.70
milliwatt (mW)	watt (W)	10^{-3}
(Pressure, stress)		
atmosphere (std) (14.7 lbf/in^2)	newton/meter^2 (N/m^2 or Pa)	101,330
pound/foot^2 (lbf/ft^2)	newton/meter^2 (N/m^2 or Pa)	47.880
pound/inch^2 (lbf/in^2 or psi)	newton/meter^2 (N/m^2 or Pa)	6,894.8
(Temperature)		
degree Fahrenheit (°F)	degree Celsius (°C)	°C = (°F -32) × 5 / 9
(Electrical)		
Volt (V); Ampere (A); Ohm (Ω)		

APPENDIX C SYSTEMS-AT-A-GLANCE TABLES

Name	Computer	Onboard Equipment	Accuracy-position [mm]	Accuracy-orientation [°]	Sampling Rate [Hz]	Features	Effective Range, Notes	Reference
General			0.01 %-5% of traveled distance		100-10,000 or analog	Error accumulation	Unlimited, internal, local	[Parish and Grabbe, 1993] Omnitech Robotics, Inc.
TRC Labmate	486-33MHz	Each quad-encoder pulse corresponds to 0.012 mm wheel displacement	4×4 meters bidirectional square path*: 310 mm	On smooth concrete*: 6° With ten bumps*: 8°	Very high – 1 KHz	Short wheelbase	Unlimited	[TRC] Transition Research Corp.
Cybermotion	Onboard proprietory	Drive and steer encoders	4×4 meters bidirectional square path*: 63 mm	On smooth concrete*: 1 to 3.8° With ten bumps*: 4°		Synchro-drive design		Cybermotion
Blanche	MC68020	Uses a pair of knife-edge non-load-bearing wheels for odometry						[Cox, 1991] NEC Research Institute
Model-reference adaptive motion control	386-20 MHz TRC Labmate	Wheel encoders and sonars for orientation measurements	Average after a 2×2 m square path: 20 mm	Average after 2×2 m square path: 0.5°	20 Hz	Can only compensate for systematic error	Unlimited	[Feng et al., 1994] Univ. of Michigan
Multiple robots		Two cooperative robots: one moves and one stays still and measures the motion of the moving one	Simulation: 8 mm after 100 meters movement at 2 m step			Capable of maintaining good position estimate over long distance	Unlimited	[Sugiyama, 1993] NTT Communication Science Lab.
CLAPPER: Dual-drive robot with internal correction of Odometry	486-33 MHz	Two TRC Labmates, connected by a compliant linkage; two absolute rotary encoders, one linear encoder	4×4 m square path: no bumps: 22 mm With 10 bumps¹: 44 mm	On smooth concrete*: 22° With 10 bumps*: 0.4°	25 Hz	Capable of compensating for random disturbance	Require additional robot or trailer	[Borenstein, 1994] Univ. of Michigan
UMBmark calibration for reduction of systematic odometry errors	486-33 MHz or any onboard computer	Any differential-drive mobile robot; tests here performed with TRC LabMate	4×4 ms square path: average return position error: 30-40 mm		25 Hz	Designed for reduction of systematic odometry errors; this calibration routine can be applied to any differential-drive robot, requires no special tooling or instrumentation		[Borenstein and Feng, 1995a,b, c] Univ. of Michigan
Fluxgate magnetometer				±1 - ±4°	10-1000 or analog	External, global, $100-2000 Prone to magnetic disturbance	Unlimited	[Parish and Grabble, 1993] Omnitech Robotics, Inc.

* This result is based on running the University of Michigan Benchmark (UMBmark) test for dead-reckoning accuracy. This test is described in detail in [Borenstein and Feng, 1994].

Name	Computer	Onboard Equipment	Accuracy-position [mm]	Accuracy-orientation [°]	Sampling Rate [Hz]	Features	Effective Range, Notes	Reference
Angular rate gyro (laser or optical fiber)			Very accurate models available at $1K-5K Problems are time dependent drift, and *minimum* detectable rate of rotation Gyro will not "catch" slow rotation errors	0.01 %-5% of full scale rate.	10-1000 or analog	Internal, local, $1K-20K.	Unlimited	[Parish and Grable, 1993] Omnitech Robotics, Inc.
Radar velocimeter (Doppler)			0.01 %-5% of full scale rate		100-1000 or analog	Internal, local, $1K-10K	Unlimited	[Parish and Grable, 1993] Omnitech Robotics, Inc.
Filtered/inertial sensor suite (direction gyros and accelerometer based)			0.01 %-5% of distance traveled, also time dependent drift		10-1000 or analog	Internal, local, $3K-$150K+	Unlimited	[Parish and Grable, 1993] Omnitech Robotics, Inc.
MiniRover MKI	Underwater vehicle	Fluxgate magnetic sensor		Accuracy: ±2% max. Resultion: 2°			0° - 359°	[BENTHOS] BENTHOS, Inc.
Futaba model helicopter gyro FP-G154	Output: pulse-width modulated signal			Drift: >1°/s	20 ms	$150		[TOWER]
Gyration *GyroEngine*	RS232 interface			Drift: 9°/min		$300	Unlimited	[GYRATION] Gyration, Inc.
Angular rate gyros, general (Laser or Optical Fiber)		Very accurate models available at $1K-5K Problems are time dependent drift, and *minimum* detectable rate of rotation Gyro will not "catch" slow rotation errors	0.01 %-5% of full scale rate.		10-1000 or analog	Internal, local, $1K-20K.	Unlimited	[Parish and Grable, 1993], Omnitech Robotics, Inc.
Hitachi OFG-3	RS232 interface or TTL	Originally designed for automotive navigation systems		Drift: 0.0028°/s	100 Hz		Unlimited	Komoriya and Oyama [1994], [HITACHI]
Andrew autogyro	RS232 interface	Quoted minimum detectable rotation rate: ±0.02°/s Actual minimum detectable rate limited by deadband after A/D conversion: 0.0625°/s		Drift: 0.005°/s	10 Hz	$1000	Unlimited	[ANDREW] Andrew Corporation
Complete inertial navigation system including ENV-05S Gyrostar solid state rate gyro, the START solid state gyro, one triaxial linear accelerometer and two inclinometers			Position drift rate 1 to 8 cm/s depending on the freq. of acceleration change	Gyro drift 5-15°/min. After compensation: drift 0.75°/min	100-1000 or analog	Internal, global	unlimited	[Barshan and Durrant-Whyte, 1993, 1995];[GEC]; [MURATA]
Non-Wire Guidance System for AGV's	VCC-2 vehicle control computer	Solid state gyroscope, position code reader	Position codes (landmarks)					[CONTROL] Control Engineering Company

Name	Computer	Onboard Components	Stationary Components	Accuracy - position [mm]	Accuracy - orientation [°]	Sampling rate [Hz]	Features	Effective Range	Manufacturer
CONAC (computerized opto-electronic navigation and control)	486-33 MHz	Structured opto-electronic acquisition beacon (STROAB)	Networked opto-electronic acquisition datum (NOAD)	Indoor ±1.3 mm outdoor ±5 mm	Indoor and outdoor ±0.05°	25 Hz	3-D - At least 3 NOADS for one acre. Networkable for unlim. area	Need line-of-sight for at least three NOADS	[MacLeod, 1993] (MTI)
ROBOSENSE		Scanning laser rangefinder	Retroreflective targets	System measures direction *and distance* to beacons with accuracy <0.17° and <20 mm, respectively Accuracy for robot location and orientation not specified		10-40 Hz	2-D - Measure both angle and distance to target	0.3-30 m	[SIMAN] SIMAN Sensors & Intelligent Machines Ltd.
NAMCO *LASERNET* beacon tracking system	RS-232 serial interface provided	Rotating mirror pans a near-infrared laser beam through a horizontal arc of 90°	Retroreflective targets of known dimensions	Angular accuracy is within ±0.05% with a resolution of 0.006° Accuracy for robot location and orientation not specified.		20 Hz	Derives distance from computing time of sweep over target of known width	15 meters (50 ft)	[NAMCO, 1989]
TRC beacon navigation system	6808 integrated computer, RS232 interface	Rotating mirror for scanning laser beam	Retroreflective targets, usually mounted on stand-alone poles	Resolution is 120 mm (4-3/4 in) in range and 0.125° in bearing for full 360° coverage in a horizontal plane		1 Hz	Currently limited to single work area of 80×80 ft	24.4 m (80 ft)	[TRC]
LASERNAV	64180 microcomputer	Laser scanner	Retroreflective bar codes. Up to 32 can be distinguished.	±1 in moving at 2 ft/sec; ±0.5 in stationary	±0.03°.	90 Hz	2-D - Measures only angles to reflectors	30 meters (100 ft) With active reflectors: up to 183 m	[Benayad-Cherif, 1992] and [DBIR]
BNS (beacon navigation system); 30.5 m		Optical IR detector (±10° field of view in horizontal and vertical axes)	Infrared beacon transmitter (uniquely identifiable, 128 codes)		0.3° in the ±5° central area and ±1° out to the periphery of the sensitive area	10 Hz		500 ft suitable for long corridors	[Benayad-Cherif, 1992] (Denning)
Laser scanner + corner cubes	8086	Laser scanner	Three corner cubes	LN-10: ±500 LN-20: ±20 LN-30: ±500 LN-40: ±20	LN-10: ±1° LN-20: ±0.1° LN-30: ±1° LN-40: ±0.1°	0.5 Hz		LN-10 50 m LN-20 50 m LN-30 200 m LN-40 200 m	[Nishide et al., 1986]. Tokyo Aircraft Instrument Co., Ltd.
Laser scanner + bar code		Laser scanner	Barcoded target			0.033 Hz			[Murray, 1991] Caterpillar
Magnetic markers			Magnetic markers buried under path (50 ft apart)						[Murray, 1991] Eaton-Kenway

181

Name	Computer	Onboard Components	Stationary Components	Accuracy - position [mm]	Accuracy - orientation [°]	Sampling rate [Hz]	Features	Note	Researchers &References
Three object triangulation	486-33 MHz	Computer vision system		Mean error (I) x=234, y=225 (G) x=304, y=301 (N) x=17, y=17 (C) x=35, y=35	Mean error (I) 4.75° (G) 141.48° (N) 2.41° (C) 5.18°	Mean time (I) 3048.7 (G) 3.8 (N) 33.5 (C) 4.8	Computer simulation for comparative study of four triangulation algorithms Accuracies are sensitive to landmark location	(I) Iterative Search (G) Geometric triangulation (N) Newton-Raphson (C) Circle intersection	[Cohen and Koss, 1992] Univ. of Michigan
Laser beam + corner cube	8086	Four laser transceivers (transmitter and receiver)	Two corner cube reflectors on both sides of the path	x=30 y=2		10 Hz			[Tsumura et al., 1988]
Ultrasonic beacons		Eight sonar receiver array (45° apart)	Six sonar beacons in a 12 m² space	Measured standard dev. of path error of 40 mm		150 ms			[Kleeman, 1992]
Infrared beacons		One optical infrared scanner	Infrared beacons	25 m² test area, beacons (0,0), (5,0) and (5,4); worst error = 70	±0.2°				[McGillem and Rappaport, 1988]
Laser scanner + corner cube	Z80	Laser scanner	Retro-reflector 45×45 m space, 3 reflectors at A(0,0),B(45,0), C(0,45)	Inside DABC: Mean=57,σ.=25 Outside DABC: mean=140, σ=156 On line AB or AC mean=74, σ=57	Inside DABC: mean=0.07 σ=0.06 Outside DABC: mean=0.13,σ=0.16 On line AB or AC: mean=0.12,σ=0.05				[Tsumura and Hashimoto, 1986]
Vision camera + retro-reflectors		Vision camera + light source	Retro-reflectors on the path	Path error within 10mm, at 1m/s		10 Hz			[Takeda et al., 1986]
Three target triangulation		Detector	Active beacon	100 with very noisy measurement			Optimize using all beacon data, reweighted least square criterion		[Durieu et al., 1989]
Direction measure of several identical beacons		Laser scanner	Strips of reflective tapes	At 0.3 m/s, error <2 cm At 1 m/s, is stable At 1.5 m/s, instable			Can navigate on wet rainy field, even when the drive wheels were spinning		[Larsson et al, 1994] University of Lulea
Triangulation with more than 3 landmarks			3 to 20 beacons.	6.5 cm in 10×10 m area.	Simulation results only, but simulation includes model of large measurement errors When many beacons available, system can identify and discard outliers (i.e., large errors in the measured angles to some of the beacons)				[Betke and Gurvitz, 1994], MIT

Name	Computer	Onboard Components	Features used	Accuracy - position [mm]	Accuracy - orientation [°]	Sampling Rate [Hz]	Features	Effective Range, Notes	Reference
Camera vision robot position and slippage control system	PC	Vision camera	Rectangular ceiling lights, concentric circle	<100 mm		>1 Hz			Cyberworks, Inc. [CYB]
Absolute positioning using a single image	68030, 25 MHz	Fixed vision camera (6 m high) discretization 9.5×6.0 mm for one pixel	Known pattern composed of coplanar points (IR diodes) Test pattern: 1.0×2.8 m. 84 uniformly distributed points	Accuracy: mean=2,max:10 repeatability X: mean=0.7,max: 2 σ = 0.8 Y: mean: 2 max: 5, std. 2	Repeatability mean: 0.3° max: 0.7° std. 0.4°	4 Hz	Can monitor robot operation at the same time. 3-D operation.		[Fleury and Baron, 1992] Laboratoire d'Automatique et d'Analyse des Systemes
Real-time vision-based robot localization	Sun 4/280 computer Karlsruhe mobile robot system (KAMRO)	780×580 CCD-camera, f=8 mm VISTA real-time image processing system	Vertical edges matching using stored map	15 mm	0.1°	2 Hz	Correspondence between observed landmarks and a stored map, give bond on the localization error 2-D operation		[Atiya and Hager, 1993] University of Karlsruhe
Robot localization using common object shapes	Sun workstation	640×400×4b CCD camera, PC-EYE imaging interface	Objects with a polygon-shaped top and a lateral surface perpendicular to the top	<5%			Sensitive at certain orientations		[Chen and Tsai, 1991] National Chaio Tung University
Omnidirectional vision navigation with beacon recognition		Vision camera with fish-eye lens	A light array (3x3)	40 mm	0.3°				[Cao et al., 1986] University of Cincinnati
Vision algorithm for mobile vehicle navigation	TRC Labmate	Vision camera	Two sets of four coplanar points are necessary	7 m distance 10%					[D'Orazio et al., 1991] Istituto Elaborazione Segnali ed Immagini
Adaptive position estimation	Litton S-800 486 control MC68000 positioning	Camera, strobe, landmark	Two circles of different radii	5 mm			Convergence 120 measurements	Adapt system model using maximum likelihood algorithm	[Lapin, 1992] Georgia Institute of Technology
Guidance system using optical reflectors	Sun	Camera, strobe light, (only on 0.3 s)	Reflector pattern mounted on the ceiling 2 m high						[Mesaki and Masuda, 1992] Secom Intelligent Systems Laboratory

Name	Computer	Onboard Components	Features used	Accuracy - position [mm]	Accuracy - orientation [°]	Sampling Rate [Hz]	Features	Effective Range, Notes	Reference
Positioning using a single calibrated object		Camera	A sphere with horizontal and vertical calibration great circles	5%	5°			3-D angle error increases as great circles approach the edge of the sphere Distance error increases with the distance between the robot and landmark	[Magee and Aggarwal, 1984] University of Texas
Model based vision system	TRC LabMate 68040	512×512 gray-level CCD camera, f=6 mm	Corners of the room	100 mm middle error 2%	±3°			3-D orientation error <0.5°. if the corner is in the center of the image Large error when corner is off image center and angle coefficients of L and R are too small	[D'Orazio et al., 1993] Istituto Elaborazione Segnali ed Immagini
Pose estimation	9200 image processor	Fairchild 3000 CCD camera (256×256), f=13mm Percepties	Quadrangular target s12=77.5,s13=177.5 s14=162,s23=160 s24=191,s34=104	At 1500 mm: 11 mm	At 1500 mm: 1.5°.			3-D volume measurement of tetrahedra composed of feature point triplets extracted from an arbitrary quadrangular target and the lens center	[Abidi and Chandra, 1990] University of Tennessee
Positioning using standard pattern			Relative displacement pattern: circle, half white & half black Identification pattern: bar code	At 5000 mm: 2.2%	Largest error: 2°			Errors increase with increasing distance, angle between landmark and camera too small or too large	[Kabuka and Arenas, 1987] University of Miami
TV image processing for robot positioning			Diamond shape, 90° angle and 23 cm each side	At 4000 mm: 70 mm	At 4000 mm: ±2°	90 s processing time	2-D	Errors increase with distance and angle too small or too large	[Fukui, 1981] Agency of Industrial Science and Technology
Single landmark navigation	ARCTEC Gemini robot	Infrared detector (angular resolution ±4°)	Infrared beacons	At 4000 mm: 400 mm At 2400 mm: 200 mm			2-D, error increases with the increase of distance between the vehicle and beacon	Running fix: using dead-reckoning info to use measurement obtained at t(k-1) at time t(k)	[Case, 1986] US Army Construction Eng. Research Lab.
Robot positioning using opto-electronic processor	386 PC Image-100 image processing board	256×256 camera, f=16 mm Hough transform filter (128×128)	Circle (R=107mm)	At 2000 mm 35 mm		30 Hz	2-D, the result is the fusion of dead reckoning and observed	Errors are function of the distance and angle	[Feng et al., 1992] University of Michigan
Global vision		Camera mounted at fixed points in the environment					Large range over which obstacles can be detected, allows global path planning	Main problems: how many cameras and where to put them?	[Kay and Luo, 1993] North Carolina State University

Name	Computer	Onboard Components	Features used	Accuracy - position [mm]	Accuracy - orientation [°]	Sampling Rate [Hz]	Features	Effective Range, Notes	Reference
Robot localization using a single image		Sony CCD camera, f=8.5mm resolution = 0.12°/pixel at image center	Vertically oriented parts of fixed objects, e.g., doors, desks and wall junctions Stored map		Min. distance to landmark: 1000 mm. orientation 0.2°		2-D	Utilizes the good angular resolution of a CCD camera, avoids feature correspondence and 3-D reconstruction	[Krotkov, 1991] Laboratoire d'Automatique et d'Analyse des Systemes
Autonomous robot for a known environment (ARK)	Two VME-based cards	CCD camera, IR spot laser rangefinder, custommade pan/tilt table	"Natural" landmarks, e.g., semipermanent structures, doorways)	On the order of centimeters				On the order of < 10 m.	[AECL]
Z-shaped landmark	Two onboard computer (type not specified)	Gyro-compass, odometry, metal sensor	Z-shaped, metal landmarks along the road Vehicle re-calibrates position when traveling over landmark Accuracy after re-calibration: 10 cm					Unlimited	[Matsuda and Yoshikawa, 1989], Komatsu Ltd., Tokyo, Japan.

Name	Computer and Robot	Onboard Components	Maps and Features	Accuracy - position [mm]	Accuracy - orientation [°]	Sampling Rate [Hz]	Features	Effective Range, Notes	Reference
Scanning laser rangefinder				0.5%-5%		1 to 10 kHz or analog	External, local, $10K-100K	300 m	[Parish and Grabble, 1993], Omnitech Robotics, Inc.
Scanning IR rangefinder				1%-10%		100-1000 or analog	External, local, $5K-20K	5-50 m	[Parish and Grabble, 1993], Omnitech Robotics, Inc.
Scanning (or arrayed) ultrasonic rangefinder				1%-10%		1-100	External, local, $100-5K	1-10 m	[Parish and Grabble, 1993], Omnitech Robotics, Inc.
Visual				1%-20%		0.1-100	External, local, $500-50K	1-10000	[Parish and Grabble, 1993], Omnitech Robotics, Inc.
Navigation by multi-sensory integration	TRC Labmate	Cohu CCD camera, f=16 mm dead reckoning					Integrates position estimates from vision system with odometry using Kalman filter framework		[D'Orazio et al., 1993] CNR-IESI
Laserradar and sonar based world modeling	Tricycle robot	24 sonars. four laser rangefinders, rotate at 360°/s, each scan 720 range points					Utilizes heterogeneous info from laser radar and sonars		[Buchberger et al., 1993] Kaiserslautern University
Vision directed navigation	Sun Sparc for vision, Micro-VAX as host, ROBMAC100 tricycle type vehicle	Vision camera	Doors, columns	±5.0 cm	2.0°	2 Hz Convex and concave polygons	3-D		University of Waterloo [Wong and Gao, 1992]
Robot localization by tracking geometric beacons	Sun-3 for localization Sun-4 vehicle control	One rotating sonar or six fixed sonars	Geometric beacon - naturally occurring environment feature			1 Hz	EKF utilizes matches between observed geometric beacons and a priori map of beacon locations		[Leonard and Durrant-Whyte, 1991] University of Oxford

Name	Computer and Robot	Onboard Components	Maps and Features	Accuracy - position [mm]	Accuracy - orientation [°]	Sampling Rate [Hz]	Features	Effective Range, Notes	Reference
Position estimation using vision and odometry	Differential-drive vehicle 386 PC	756×581 CCD camera f=12.5 mm	Vertical edges and stored map	40 mm	0.5°		2-D - Realistic odometry model and its uncertainty is used to detect and calculate position update fused with observation	Extended Kalman filter to correct the vehicle pose from the error between the observed and estimate angle to each landmark	[Chenavier and Crowley, 1992] LETI-DSYS
Recognize world location with stereo vision		Stereo cameras	Long, near vertical stereo features				1000 real-world data recognition test, under 10% false negative, zero false positive	Least-squares to find the best fit of model to data and evaluate that fit	[Brauuegg, 1993] MITRE Corp.
Environment learning using a distributed representation	Omnidirectional three-wheeled base	a ring of 12 sonars and a compass	Left wall, right wall, corridors				Dynamic landmark detection utilizing robot's motion	Learn the large-space structure of environment by recording its permanent features	[Mataric, 1990] MIT
Localization in structured environment	Motorola M68020	A ring of 24 sonars	Classify objects into edges, corners, walls, and unknown objects			0.1 Hz	Positions resulting from all possible mappings are calculated and then analyzed for clusters The biggest cluster is assumed to be at the true robot position	Each mapping of two model objects onto two reference objects correspond to a certain robot position	[Holenstein et al., 1992] Swiss Federal Inst. of Technology
Localization using sonar	SUN 4	Linear array of three sonars: A. reduce the angular uncertainty, B. help identify the target's class	Local map: feature map (extended reflectors, e.g., wall, and point reflectors)	<10 mm	<1°		Local map: feature extraction Matching: least squares EKF for estimating the geometric parameters of different targets and related uncertainty	[Sabatini and Benedetto, 1994] Scuola Superiore di Studi Universitari
Sonar-based real-world mapping	Neptune mobile robot	Sonars	Probability based occupancy grid map	Map with 3000 6 in cells made from 200 well spaced readings of a cluttered 20×20 ft room can be matched with 6 in displacement and 3° rotation in 1 s of VAX time			Map matching by convolving them It gives the displacement and rotation that best brings one map into registration with the other, with a measure of the goodness of match		[Elfes, 1987] Carnegie-Mellon University

Name	Computer and Robot	Onboard Components	Maps and Features	Accuracy - position [mm]	Accuracy - orientation [°]	Sampling Rate [Hz]	Features	Effective Range, Notes	Reference
Comparison of grid-type map building by index of performance (IOP)	Cybermotion K2A synchro-drive robot 386 20 MHz PC	A ring of 24 sonars	Histogramic in-motion mapping (HIMM) and heuristic probability function	HIMM results in a sensor grid in which entries in close proximity to actual object locations have a a favorable (low) Index of performance value			Index of performance (IOP) computes the correlation between the sensed position of objects, as computed by the map-building algorithm, and the actual object position, as measured manually The IOP gives quantitative measure of the differences in the sensor grid maps produced by each algorithm type		[Raschke and Borenstein, 1990] University of Michigan
Comparison of position estimation using occupancy grid			Local map: grid map Global map: grid map	Best result obtained by matching segment to segment			Grid to segment matching: generating a mask for the segment and correlating it with the grid map	Segment to segment matching: A. orientation B. collinearity C. overlap	[Schiele and Crowley, 1994] LIFIA
Blanche	MC68020 Tricycle-type mobile robot	Optical range-finder, res. = 1 in at 5 ft, 1000 samples/rev. in one s. Odometry	24 line-segment environments map for a 300×200 in room	6 in path following		Position update every 8 s for a 180 points image and a map of 24 lines. 2-D map.	(1) Least-square for data and model matching (2) Combine odometry and matching for better position estimate using maximum likelihood	Segments Assume the displacement between the data and model is small	[Cox, 1991] NEC Research Institute
Range map pose estimation	SPARC1+	1-D Laser range finder 1000 points/rev	Line segment, corner	Mean error Feature-based: 60 Iconic estimator: 40 In a 10×10 m space	Max under 1.2° for both	Feature-based: 0.32 s Iconic: 2 s	1000 points/rev. Iconic approach matches every range data point to the map rather than condensing data into a small set of features to be matched to the map		[Schaffer et al., 1992] CMU
Positioning using model-based maps		A rotatable ring of 12 polaroid sonars	Line segments	3-5 cm Coverge if initial estimate is within 1 meters of the true position			Classification of data points Weighted voting of correction vectors	Clustering sensor data points. Line fitting.	[MacKenzie and Dudek, 1994] McGill University
Positioning using optical range data	INMOS-T805 transputer	Infrared scanner	Line segment	The variance never exceeds 6 cm			Kalman filter position estimation Line fitting Matching, only good matches are accepted	When scans were taken from errorrous pos. matches consistently fail	[Borthwick et al., 1994] University of Oxford

Name	Computer and Robot	Onboard Components	Maps and Features	Accuracy - position [mm]	Accuracy - orientation [°]	Sampling Rate [Hz]	Features	Effective Range, Notes	Reference
World modeling and localization using sonar ranging		A ring of 24 sonars	Line segments	x = 33 mm covariance: 1 y = 17 mm covariance: 1	0.20° covariance: 17.106	A model for the uncertainty in sonars, and the projection of range measurement into external Cartesian coordinate	Extracting line segments from adjacent collinear range measurements and matching these line segments to a stored model	Matching includes: orientation, collinearity, and overlap by comparing one of the parameters in segment representation	[Crowley, 1989] LIFIT(IMAG)
2-D laser rangefinder map building	Sun Sparc	Cyclone 2-D laser rangefinder accuracy ±20 cm, range 50 m	Local map: line segment map Global map: line segments	Max. 5 cm average 3.8 cm		On SUN Sparc, 80 ms for local map building and 135 ms for global map update	Matching: remove segment already in the global map from local map, add new segment	Local map: clustering clustering segmentation line fitting	[Gonzalez et al., 1994] Universidad de malaga
Iconic position estimator	Locomotion emulator, all-wheel drive and all-wheel steer, Sun Sparc 1	Cyclone laser range scanner, resolution = 10 cm range = 50m 1000 readings per rev.	In general, has a large number of short line segments	Max. 36 mm mean 19.9 mm	Max. 1.8° mean 0.73°		Iconic method works directly on the raw sensed data, minimizing the discrepancy between it and the model	Assume small displacement between sensed data and model Two parts: sensor to map data correspondence & error minimization	[Gonzalez et al., 1992] Carnegie Mellon University
Environment representation from image data			Geometrical relationships between observed features rather than their absolute position				A graph where the nodes represent the observed features and edges represent the relationships between features	The recognition problem can be formulated as a graph matching problem	[Taylor, 1991] Yale University
Localization via classification of multi-sensor maps		Sonars Lateral motion vision Infrared proximity sensor	Local map: multi-sensor 100×100 grid maps, cell 20×20 cm	Using datasets from 10 rooms and hallways, estimate a 94% recognition rate for rooms, and 98% for hallways		Local grid maps Feature-level sensor fusion by extracting spatial descriptions from these maps	Positioning by classifying the map descriptions to recognize the workspace region that a given map represents	Matching: K-nearest neighbor and minimum Mahalanobis distance	[Courtney and Jain, 1994] Texas Instruments, Inc.

Name	Computer	Onboard Components	Maps and Features	Accuracy - position [mm]	Accuracy - orientation [°]	Sampling Rate [Hz]	Features	Effective Range, Nnotes	Reference
Guide path sensor (magnetic, optical, inductive, etc.)				0.01-0.1 m		100-1000 or analog	External, local, or waypoint indication, $100-$5K	0.01-0.2 m	[Parish and Grabble, 1993] Omnitech Robotics, Inc.
Odor trails for navigation		Applicator for laying volatile chemicals on the floor; olfactory sensor						Unlimited	[Russell et al., 1994] Monash University
Thermal path following		Quartz halogen bulb and pyroelectric sensor				0.833	No need to remove markers after use	Unlimited	[Kleeman and Russell, 1993] Monash University

References

Subject Index

Author Index

University of Michigan grad. student Ulrich Raschke verifies the proper alignment of ultrasonic sensors. All three robots in this picture use 15°-angular spacing between the sensors. Many researchers agree that 15° spacing assures complete coverage of the area around the robot.

REFERENCES

Abidi, M. and Chandra, T., 1990, "Pose Estimation for Camera Calibration and Landmark Tracking." *Proceedings of IEEE International Conference on Robotics and Automation*, Cincinnati, OH, May 13-18, pp. 420-426.

Abidi, M. and Gonzalez, R., Editors, 1992, *Data Fusion in Robotics and Machine Intelligence*. Academic Press Inc., San Diego, CA.

Acuna, M.H. and Pellerin, C.J., 1969, "A Miniature Two-Axis Fluxgate Magnetometer." *IEEE Transactions on Geoscience Electronics*, Vol. GE-7, pp 252-260.

Adams, M.D., 1992, "Optical Range Data Analysis for Stable Target Pursuit in Mobile Robotics." Ph.D. Thesis, Robotics Research Group, University of Oxford, U.K.

Adams, M. et al., 1994, "Control and Localisation of a Post Distributing Mobile Robot." *1994 International Conference on Intelligent Robots and Systems (IROS '94)*, Munich, Germany, Sept. 12-16, pp. 150-156.

Adams, M., 1995, "A 3-D Imaging Scanner for Mobile Robot Navigation." Personal Communication. Contact: Dr. Martin Adams, Institute of Robotics, Leonhardstrasse 27, ETH Centre, CH-8092, Switzerland. Ph.: +41-1-632-2539. E-mail: adams@ifr.ethz.ch.

Adams, M.D. and Probert, P.J., 1995, "The Interpretation of Phase and Intensity Data from A.M.C.W. Light Detection Sensors for Reliable Ranging." Accepted for publication in the *IEEE International Journal of Robotics Research*, April.

Adrian, P., 1991, "Technical Advances in Fiber-Optic Sensors: Theory and Applications." *Sensors*, Sept.pp. 23-45.

Agent, A., 1991, "The Advantages of Absolute Encoders for Motion Control." *Sensors*, April, pp. 19-24.

Allen, D., Bennett, S.M., Brunner, J., and Dyott, R.B., 1994, "A Low Cost Fiber-optic Gyro for Land Navigation." Presented at the SPIE Annual Meeting, San Diego, CA, July.

Arkin, R.C., 1989, "Motor-Schema-Based Mobile Robot Navigation." *International Journal of Robotics Research*, Vol. 8., No. 4, Aug., pp. 92-112.

Aronowitz, F., 1971, "The Ring Laser Gyro," *Laser Applications*, Vol. 1, M. Ross, ed., Academic Press.

Arradondo-Perry, J., 1992, "GPS World Receiver Survey." *GPS World*, January, pp. 46-58.

Atiya, S. and Hager, G., 1993, "Real-time Vision-based Robot Localization." *IEEE Transactions on Robotics and Automation*, Vol. 9, No. 6, pp. 785-800.

Aviles, W.A. et al., 1991, "Issues in Mobile Robotics: The Unmanned Ground Vehicle Program Teleoperated Vehicle (TOV)." *Proceedings of the SPIE - The International Society for Optical Engineering*, Vol: 1388 p. 587-97.

Avolio, G., 1993, "Principles of Rotary Optical Encoders." *Sensors*, April, pp. 10-18.

Baines, N. et al., 1994, "Mobile Robot for Hazardous Environments." Unpublished paper. Atomic Energy of Canada, Ltd., Sheridan Research Park, 2251 Speakman Drive, Mississauga, Ontario, L5K 1B2, Canada, 416-823-9040.

Baker, A., 1993, "Navigation System Delivers Precision Robot Control." *Design News*, Dec. 20, p. 44.

Banzil, G., et. al., 1981, "A Navigation Subsystem Using Ultrasonic Sensors for the Mobile Robot Hilare." Proceedings of 1st Conference on Robot Vision and Sensory Control, Stratford/Avon, U.K., April 13.

Barrett, C.R., Nix, W.D., and Tetelman, A.S., 1973, *"The Principles of Engineering Materials."* Prentice Hall, Englewood Cliffs, NJ.

Barshan, B. and Durrant-Whyte, H.F., 1993, "An Inertial Navigation System for a Mobile Robot." *Proceedings of the 1993 IEEE/RSJ International Conference on Intelligent Robotics and Systems*, Yokohama, Japan, July 26-30, pp. 2243-2248.

Barshan, B. and Durrant-Whyte, H.F., 1994, "Orientation Estimate for Mobile Robots Using Gyroscopic Information." *1994 International Conference on Intelligent Robots and Systems (IROS '94)*. Munich, Germany, Sept. 12-16, pp. 1867-1874.

Barshan, B. and Durrant-Whyte, H.F., 1995, "Inertial Navigation Systems Mobile Robots." *IEEE Transactions on Robotics and Automation*, Vol. 11, No. 3, June, pp. 328-342.

Benayad-Cherif, F., Maddox, J., and Muller, L., 1992, "Mobile Robot Navigation Sensors." *Proceedings of the 1992 SPIE Conference on Mobile Robots*, Boston, MA, Nov. 18-20, pp. 378-387.

Bennett, S. and Emge, S.R., 1994, "Fiber-optic Rate Gyro for Land Navigation and Platform Stabilization." Presented at *Sensors Expo '94*, Cleveland, OH, Sept. 20.

Betke, M and Gurvits, L., 1994, "Mobile Robot Localization Using Landmarks." *1994 International Conference on Intelligent Robots and Systems (IROS'94)*. Munich, Germany, Sept. 12-16, pp.135-142.

Beyer, J., Jacobus, C., and Pont, F., 1987, "Autonomous Vehicle Guidance Using Laser Range Imagery." *SPIE* Vol. 852, Mobile Robots II, Cambridge, MA, Nov, pp. 34-43.

Biber, C., Ellin, S., and Shenk, E., 1987, "The Polaroid Ultrasonic Ranging System." *Audio Engineering Society*, 67th Convention, New York, NY, Oct.-Nov.

Binger, N. and Harris, S.J., 1987, "Applications of Laser Radar Technology." *Sensors*, April, pp. 42-44.

Boltinghouse, S., Burke, J., and Ho, D., 1990, "Implementation of a 3D Laser Imager Based Robot Navigation System with Location Identification." SPIE Vol. 1388, Mobile Robots V, Boston, MA, Nov., pp. 14-29.

Boltinghouse, S. and Larsen, T., 1989, "Navigation of Mobile Robotic Systems Employing a 3D Laser Imaging Radar." *ANS Third Topical Meeting on Robotics and Remote Systems*, Section 2-5, Charleston, SC, March, pp. 1-7.

Bolz, R.E. and Tuve, G.L., Ed., 1979, *CRC Handbook of Tables for Applied Engineering Science*, CRC Press, Boca Raton, FL.

Borenstein, J. and Koren, Y., 1985, "A Mobile Platform For Nursing Robots." *IEEE Transactions on Industrial Electronics*, Vol. 32, No. 2, pp. 158-165.

Borenstein, J. and Koren, Y., 1986, "Hierarchical Computer System for Autonomous Vehicle." *Proceedings of the 8th Israeli Convention on CAD/CAM and Robotics*, Tel-Aviv, Israel, December 2-4.

Borenstein, J., 1987, "The Nursing Robot System." *Ph. D. Thesis, Technion*, Haifa, Israel, June, pp. 146-158.

Borenstein, J. and Koren, Y., 1987, "Motion Control Analysis of a Mobile Robot." *Transactions of ASME, Journal of Dynamics, Measurement and Control*, Vol. 109, No. 2, pp. 73-79.

Borenstein, J. and Koren, Y., 1990, "Real-Time Obstacle Avoidance for Fast Mobile Robots in Cluttered Environments." *IEEE International Conference on Robotics and Automation*, Vol. CH2876-1, Cincinnati, OH, pp. 572-577, May.

Borenstein, J. and Koren, Y., 1991, "The Vector Field Histogram – Fast Obstacle-Avoidance for Mobile Robots." *IEEE Journal of Robotics and Automation*, Vol. 7, No. 3., June, pp. 278-288.

Borenstein, J., 1992, "Compliant-linkage Kinematic Design for Multi-degree-of-freedom Mobile Robots ." *Proceedings of the SPIE Symposium on Advances in Intelligent Systems, Mobile Robots VII*, Boston, MA, Nov. 15-20, pp. 344-351.

Borenstein, J., 1993, "Multi-layered Control of a Four-Degree-of-Freedom Mobile Robot With Compliant Linkage." *Proceedings of the 1993 IEEE International Conference on Robotics and Automation*, Atlanta, GA, May 2-7, pp. 3.7-3.12.

Borenstein, J., 1994a, "The CLAPPER: A Dual-drive Mobile Robot with Internal Correction of Dead-reckoning Errors." *Proceedings of IEEE International Conference on Robotics and Automation*, San Diego, CA, May 8-13, pp. 3085-3090.

Borenstein, J., 1994b, "Internal Correction of Dead-reckoning Errors With the Smart Encoder Trailer." *1994 International Conference on Intelligent Robots and Systems (IROS '94)*. Munich, Germany, Sept. 12-16, pp. 127-134.

Borenstein, J., 1994c, "Four-Degree-of-Freedom Redundant Drive Vehicle With Compliant Linkage." *Video Proceedings of the 1994 IEEE International Conference on Robotics and Automation*, San Diego, CA, May 8-13.

Borenstein, J. and Feng, L., 1994, "*UMBmark* — A Method for Measuring, Comparing, and Correcting Dead-reckoning Errors in Mobile Robots." *Technical Report, The University of Michigan UM-MEAM*-94-22, Dec.

Borenstein, J., 1995, "Control and Kinematic Design for Multi-degree-of-freedom Mobile Robots With Compliant Linkage." *IEEE Transactions on Robotics and Automation*, Vol. 11, No. 1, Feb., pp. 21-35.

Borenstein, J. and Koren, Y., 1995, "Error Eliminating Rapid Ultrasonic Firing for Mobile Robot Obstacle Avoidance." *IEEE Transactions on Robotics and Automation*, Vol. 11, No. 1, Feb., pp 132-138.

Borenstein, J., Wehe, D., Feng, C., and Koren, Y., 1995, "Mobile Robot Navigation in Narrow Aisles with Ultrasonic Sensors." Presented at the *ANS 6th Topical Meeting on Robotics and Remote Systems*, Monterey, CA, Feb. 5-10.

Borenstein, J. and Feng, L., 1995a, "Measurement and Correction of Systematic Odometry Errors in Mobile Robots." Accepted for publication as a regular paper in the *IEEE Transactions on Robotics and Automation*, Apr.

Borenstein, J. and Feng. L., 1995b, "Correction of Systematic Dead-reckoning Errors in Mobile Robots." *Proceedings of the 1995 International Conference on Intelligent Robots and Systems (IROS '95)*, Pittsburgh, PA, Aug. 5-9, pp. 569-574.

Borenstein, J. and Feng. L., 1995c, "UMBmark: A Benchmark Test for Measuring Dead-reckoning Errors in Mobile Robots." Accepted for presentation at the *1995 SPIE Conference on Mobile Robots*, Philadelphia, October 22-26.

Borenstein, J., 1995, Video, "The CLAPPER: A Dual-drive Mobile Robot With Internal Correction of Dead-reckoning Errors." *Video Proceedings of the 1995 IEEE International Conference on Robotics and Automation,* Nagoya, Japan, May 21-27.

Brooks, R., 1985, "Visual Map Making for a Mobile Robot." *Proceedings of IEEE International Conference on Robotics and Automation*, St. Louis, MO, March 25-28, pp. 824-829.

Brown, R.G., Hwang, P.Y.C., 1992, *Introduction to Random Signals and Applied Kalman Filtering.* 2nd ed., John Wiley and Sons, New York, NY.

Buchberger, M., Jörg, K., and Puttkamer, E., 1993, "Laserradar and Sonar Based World Modeling and Motion Control for Fast Obstacle Avoidance of the Autonomous Mobile Robot MOBOT-IV." *Proceedings of IEEE International Conference on Robotics and Automation*, Atlanta, GA, May 10-15, pp. 534-540.

Buholz, N. and Chodorow, M., 1967, "Acoustic Wave Amplitude Modulation of a Multimode Ring Laser." *IEEE Journal of Quantum Electronics*, Vol. QE-3, No. 11, Nov., pp. 454-459.

Bulkeley, D., 1993, "The Quest for Collision-Free Travel." *Design News*, Oct. 4.

Burns, W.K., Chen, C.L., and Moeller, R.P., 1983, "Fiber-Optic Gyroscopes with Broad-Band Sources." *IEEE Journal of Lightwave Technology*, Vol. LT-1, p. 98.

Byrd, J.S. and DeVries, K.R., 1990, "A Six-Legged Telerobot for Nuclear Applications Development, *International Journal of Robotics Research*, Vol. 9, April, pp. 43-52.

Byrne, R.H., 1993, "Global Positioning System Receiver Evaluation Results." *Sandia Report SAND93-0827*, Sandia National Laboratories, Albuquerque, NM, Sept.

Byrne, R.H., Klarer, P.R., and Pletta, J.B., 1992, "Techniques for Autonomous Navigation." *Sandia Report SAND92-0457*, Sandia National Laboratories, Albuquerque, NM, March.

Cao, Z., Roning, J., and Hall, E., 1986, "Omnidirectional Vision Navigation Integrating Beacon Recognition with Positioning." *Proceedings of the 1986 SPIE Conference on Mobile Robots,* Cambridge, MA, Oct. 30-31, pp. 213-220.

Carter, E.F., Ed., 1966, *Dictionary of Inventions and Discoveries*, Crane, Russak, and Co., New York, NY.

Case, M., 1986, "Single Landmark Navigation by Mobile Robot." *Proceedings of the 1986 SPIE Conference on Mobile Robots,* Cambridge, MA, Oct. 30-31, pp. 231-237.

Chao, S., Lim, W.L., and Hammond, J.A., 1984, "Lock-in Growth in a Ring Laser Gyro." *Proceedings, Physics and Optical Ring Gyros Conference*, SPIE Vol 487, Snowbird, UT, January, pp. 50-57.

Chen, S. and Tsai, W., 1991, "Determination of Robot Locations by Common Object Shapes." *IEEE Transactions on Robotics and Automation*, Vol. 7, No. 1, pp. 149-156.

Chen, Y.D., Ni, J., and Wu, S.M., 1993, "Dynamic Calibration and Compensation of a 3D Lasar Radar Scanning System." *IEEE International Conference on Robotics and Automation*, Atlanta, GA, Vol. 3, May, pp. 652-664.

Chenavier, F. and Crowley, J., 1992, "Position Estimation for a Mobile Robot Using Vision and Odometry." *Proceedings of IEEE International Conference on Robotics and Automation*, Nice, France, May 12-14, pp. 2588-2593.

Chesnoy, J., 1989, "Picosecond Gyrolaser." *Optics Letters*, Vol 14, No. 18, Sept., pp. 990-992.

Chow, W.W., Gea-Banacloche, J., Pedrotti, L.M., Sanders, V.E., Schleich, W., and Scully, M.O., 1985, "The Ring Laser Gyro." *Reviews of Modern Physics*, Vol. 57, No. 1, January, pp. 61-104.

Christian, W.R.and Rosker, M.J., 1991, "Picosecond Pulsed Diode Ring Laser Gyroscope." *Optics Letters*, Vol. 16, No. 20, Oct., pp. 1587-1589.

Clark, R.R., 1994, "A Laser Distance Measurement Sensor for Industry and Robotics." *Sensors*, June, pp. 43-50.

Cohen, C. and Koss, F., 1992, "A Comprehensive Study of Three Object Triangulation." *Proceedings of the 1993 SPIE Conference on Mobile Robots,* Boston, MA, Nov. 18-20, pp. 95-106.

Congdon, C., et al, 1993, CARMEL Versus FLAKEY — A Comparison of Two Winners. *AI Magazine*, Winter, pp. 49-56.

Conrad, D.J. and Sampson, R.E., 1990, "3D Range Imaging Sensors." in *Traditional and Non-Traditional Robotic Sensors*, T.C. Henderson, ed., *NATO ASI Series*, Vol. F63, Springer-Verlag, pp. 35-47.

Cooper, S. and Durrant-Whyte, H., 1994, "A Kalman Filter for GPS Navigation of Land Vehicles." *1994 International Conference on Intelligent Robots and Systems (IROS '94)*. Munich, Germany, Sept. 12-16, pp. 157-163.

Courtney, J. and Jain, A., 1994, "Mobile Robot Localization via Classification of Multisensor Maps." *Proceedings of IEEE International Conference on Robotics and Automation*, San Diego, CA, May 8-13, pp. 1672-1678.

Cox, I.J., 1991, "Blanche - An Experiment in Guidance and Navigation of an Autonomous Mobile Robot." *IEEE Transactions Robotics and Automation*, 7(3), pp. 193-204.

Crowley, J., 1989, "World Modeling and Position Estimation for a Mobile Robot Using Ultrasonic Ranging." *Proceedings of IEEE International Conference on Robotics and Automation*, Scottsdale, AZ, May 14-19, pp. 674-680.

Crowley, J.L. and Reignier, P., 1992, "Asynchronous Control of Rotation and Translation for a Robot Vehicle." *Robotics and Autonomous Systems*, Vol. 10, pp. 243-251.

DeCorte, C., 1994, "Robots Train for Security Surveillance." *Access Control*, June, pp. 37-38.

Deveza, R., Thiel, D., Russell, R.A., and Mackay-Sim, A., 1994, "Odour Sensing for Robot Guidance." *The International Journal of Robotics Research,* Vol. 13, No. 3, June, pp. 232-239.

D'Orazio, T., Capozzo, L., Ianigro, M., and Distante, A., 1993, "Model Based Vision System for Mobile Robot Position Estimation." *Proceedings of the 1993 SPIE Conference on Mobile Robots,* Boston, MA, Sept. 9-10, pp. 38-49.

D'Orazio, T., Distante, A., Attolico, G., Caponetti, L., and Stella, E., 1991, "A Vision Algorithm for Mobile Vehicle Navigation." *Proceedings of the 1991 SPIE Conference on Mobile Robots,* Boston, MA, Nov. 14-15, pp. 302-309.

D'Orazio, T., Ianigro, M., Stella, E., Lovergine, F., and Distante, A., 1993, "Mobile Robot Navigation by Multi-Sensory Integration." *Proceedings of IEEE International Conference on Robotics and Automation*, Atlanta, GA, May 10-15, pp. 373-379.

Dahlin, T. and Krantz, D., 1988, "Low-Cost, Medium-Accuracy Land Navigation System." *Sensors*, Feb., pp. 26-34.

Depkovich, T. and Wolfe, W., 1984, "Definition of Requirements and Components for a Robotic Locating System." *Final Report No. MCR-83-669*, Martin Marietta Aerospace, Denver, CO, February.

Dibburn, U. and Petersen, A., 1983, "The Magnetoresistive Sensor - A Sensitive Device for Detecting Magnetic Field Variations." *Electronic Components and Applications*, Vol. 5, No. 3, June.

Dodington, S.H., 1989, "Electronic Navigation Systems." *Electronic Engineer's Handbook,* D. Christiansen and D. Fink, eds., 3rd edition, McGraw Hill, New York, pp. 76-95.

Dunlap, G.D. and Shufeldt, H.H., *Dutton's Navigation and Piloting*, Naval Institute Press, pp. 557-579.

Durieu, C., Clergeot, H., and Monteil, F., 1989, "Localization of a Mobile Robot with Beacons Taking Erroneous Data Into Account." *Proceedings of IEEE International Conference on Robotics and Automation*, Scottsdale, AZ, May 14-19, pp. 1062-1068.

Duchnowski, L.J., 1992, "Vehicle and Driver Analysis with Real-Time Precision Location Techniques." *Sensors*, May, pp. 40-47.

Edlinger, T. and Puttkamer, E., 1994, "Exploration of an Indoor Environment by an Autonomous Mobile Robot." *International Conference on Intelligent Robots and Systems (IROS '94)*. Munich, Germany, Sept. 12-16, pp. 1278-1284.

Elfes, A., 1987, "Sonar-Based Real-World Mapping and Navigation." *IEEE Journal of Robotics and Automation*, Vol. RA-3, No. 3, pp. 249-265.

Elfes, A., 1989, "Using Occupancy Grids for Mobile Robot Perception and Navigation." *Computer*, June, pp. 46-57.

Ellowitz, H.I., 1992, "The Global Positioning System." *Microwave Journal*, April, pp. 24-33.

Engelson, S. and McDermott, D., 1992, "Error Correction in Mobile Robot Map Learning." *Proceedings of IEEE International Conference on Robotics and Automation*, Nice, France, May 12-14, pp. 2555-2560.

Evans, J. M., 1994, "HelpMate: An Autonomous Mobile Robot Courier for Hospitals." *1994 International Conference on Intelligent Robots and Systems (IROS '94)*. Munich, Germany, Sept. 12-16, pp. 1695-1700.

Everett, H.R., 1982, "A Computer Controlled Autonomous Sentry Robot." Masters Thesis, Naval Postgraduate School, Monterey, CA, October.

Everett, H.R., 1985, "A Multi-Element Ultrasonic Ranging Array." *Robotics Age*, July, pp. 13-20.

Everett, H.R., Gilbreth, G.A., Tran, T., and Nieusma, J.M., 1990, "Modeling the Environment of a Mobile Security Robot." *Technical Document 1835, Naval Command Control and Ocean Surveillance Center*, San Diego, CA, June.

Everett, H.R., Gage, D.W., Gilbreth, G.A., Laird, R.T., and Smurlo, R.P., 1994, "Real-World Issues in Warehouse Navigation." *Proceedings SPIE Mobile Robots IX*, Volume 2352, Boston, MA, Nov.2-4.

Everett, H. R., 1995, *Sensors for Mobile Robots: Theory and Application*, ISBN 1-56881-048-2, A K Peters, Ltd., Wellesley, MA.

Ezekial, S. and Arditty, H.J., Ed., "Fiber Optic Rotation Sensors and Related Technologies." *Proceedings of the First International Conference*, MIT, Springer-Verlag, New York.

Fan, Z., Borenstein, J., Wehe, D., and Koren, Y., 1994, "Experimental Evaluation of an Encoder Trailer for Dead-reckoning in Tracked Mobile Robots." *Technical Report, The University of Michigan*, UM-MEAM-94-24, December.

Fan, Z., Borenstein, J., Wehe, D., and Koren, Y.,1995, "Experimental Evaluation of an Encoder Trailer for Dead-reckoning in Tracked Mobile Robots" To be presented at the *10th IEEE International Symposium on Intelligent Control*, Aug. 27-29.

Feng, L., Koren, Y., and Borenstein, J., 1994, "A Model-Reference Adaptive Motion Controller for a Differential-Drive Mobile Robot." *Proceedings of IEEE International Conference on Robotics and Automation*, San Diego, CA, May 8-13, pp. 3091-3096.

Feng, L., Fainman, Y., and Koren, Y., 1992, "Estimate of Absolute Position of Mobile Systems by Opto-electronic Processor," *IEEE Transactions on Man, Machine and Cybernetics*, Vol. 22, No. 5, pp. 954-963.

Fenn, R.C., Gerver, M.J., Hockney, R.L., and Johnson, B.G., 1992, "Microfabricated Magnetometer Using Young's Modulous Changes in Magnetoelastic Materials." *SPIE* Vol. 1694.

Figueroa, J.F. and Lamancusa, J.S., 1992, "A Method for Accurate Detection of Time of Arrival: Analysis and Design of an Ultrasonic Ranging System." Journal of the Acoustical Society of America, Vol. 91, No. 1, January, pp. 486-494.

Figueroa, J.F., Doussis, E., and Barbieri, E., 1992, "Ultrasonic Ranging System for 3-D Tracking of a Moving Target." 92-WA/DSC-3, Proceedings, Winter Annual Meeting, American Society of Mechanical Engineers, Anaheim, CA, November.

Figueroa, J.F. and Barbieri, E., 1991, "Increased Measurement Range Via Frequency Division in Ultrasonic Phase Detection Methods." *Acustica*, Vol. 73, pp. 47-49.

Figueroa, J.F. and Mahajan, A., 1994, "A Robust Navigation System for Autonomous Vehicles Using Ultrasonics." *Control Engineering Practice*, Vol. 2, No. 1, pp. 49-59.

Fisher, D., Holland, J.M., and Kennedy, K.F., 1994, "K3A Marks Third Generation Synchro-Drive." *American Nuclear Society Winter Meeting, Proceedings of Robotics and Remote Systems*, New Orleans, LA, June.

Fleury, S. and Baron, T., 1992, "Absolute External Mobile Robot Localization Using a Single Image." *Proceedings of the 1992 SPIE Conference on Mobile Robots,* Boston, MA, Nov. 18-20, pp. 131-143.

Fox, K., 1993, "Indoor Robots Starts Flying Blind." *Science,* Vol. 261, Aug. 6, pp. 685.

Fraden, J., 1993, *AIP Handbook of Modern Sensors*, ed., Radebaugh, R., American Institute of Physics, New York.

Frederiksen, T.M. and Howard, W.M., 1974, "A Single-Chip Monolithic Sonar System." *IEEE Journal of Solid State Circuits,* Vol. SC-9, No. 6, December.

Fukui, I., 1981, "TV Image Processing to Determine the Position of a Robot Vehicle." *Pattern Recognition,* Vol. 14, pp. 101-109.

Getting, I.A., 1993, "The Global Positioning System," *IEE Spectrum,* December, pp. 36-47.

Geyger, W.A., 1957, *Magnetic Amplifier Circuits*, 2nd ed., McGraw-Hill, New York.

Gilbert, W., 1992, "De Magnete." 1600. (Translation: P.F. Mottelay, John Wiley, 1893.)

Gonzalez, J., Stentz, A., and Ollero, A., 1992, "An Iconic Position Estimator for a 2D Laser RangeFinder." *Proceedings of IEEE International Conference on Robotics and Automation,* Nice, France, May 12-14, pp. 2646-2651.

Gonzalez, R. and Wintz, P., 1977, "*Digital Image Processing.*" Addison-Wesley, Reading, MA.

Gonzalez, J., Ollero, A., and Reina, A., 1994, "Map Building for a Mobile Robot Equipped with a 2D Laser Rangefinder." *Proceedings of IEEE International Conference on Robotics and Automation,* San Diego, CA, May 8-13, pp. 1904-1909.

Gothard, B.M., Etersky, R.D., and Ewing, R.E., 1993, "Lessons Learned on a Low-Cost Global Navigation System for the Surrogate Semi-Autonomous Vehicle." *SPIE Proceedings*, Vol. 2058, Mobile Robots VIII, pp. 258-269.

Gould, L., 1990, "Is Off-Wire Guidance Alive or Dead?" *Managing Automation,* May, pp. 38-40.

Gourley, C. and Trivedi, M., 1994, "Sensor Based Obstacle Avoidance and Mapping for Fast Mobile Robots." *Proceedings of IEEE International Conference on Robotics and Automation,* San Diego, CA, May 8-13, pp. 1306-1311.

Grenoble, B., 1990, "Sensor and Logic Form Digital Compass." *Electronic Design News,* Dec. 6, pp. 228-229.

Gunther, J., 1994, "Robot Asks, Where Am I?" *Popular Science,* Feb., pp. 32.

Hammond, W., 1993, "Smart Collision Avoidance Sonar Surpasses Conventional Systems." *Industrial Vehicle Technology '93: Annual Review of Industrial Vehicle Design and Engineering*, UK and International Press, pp. 64-66.

Harmon, S.Y., 1986, "USMC Ground Surveillance Robot (GSR): Lessons Learned." *Mobile Robots*, SPIE Vol. 727, Cambridge, MA, pp. 336-343.

Harris, J.C., 1994, "An Infogeometric Approach to Telerobotics," Proceedings, IEEE National Telesystems Conference, San Diego, CA, May, pp. 153- 156.

Henkel, S.L., 1987, "Optical Encoders: A Review." *Sensors,* September, pp. 9-12.

Henkel, S.L., 1994, "GMR Materials Advance Magnetic Field Detection." *Sensors,* June, p.8.

Hine, A., 1968, *Magnetic Compasses and Magnetometers*, Adam Hilger Ltd., London.

Hinkel, R. and Knieriemen, T., 1988, "Environment Perception with a Laser Radar in a Fast Moving Robot." *Symposium on Robot Control 1988 (SYROCO '88),* Karlsruhe, Germany, October 5-7, pp. 68.1 - 68.7.

Holenstein, A., Muller, M., and Badreddin, E., 1992, "Mobile Robot Localization in a Structured Environment Cluttered with Obstacles." *Proceedings of IEEE International Conference on Robotics and Automation,* Nice, France, May 12-14, pp. 2576-2581.

Holland, J.M., 1983, *Basic Robotics Concepts*, Howard W. Sams, Macmillan, Inc., Indianapolis, IN.

Holle, S., 1990, "Incremental Encoder Basics." *Sensors,* April, pp. 22-30.

Hollingum, J., 1991, "Caterpillar make the earth move: automatically." *The Industrial Robot,* Vol. 18, No. 2, pp. 15-18.

Hongo, T., Arakawa, H., Sugimoto, G., Tange, K., and Yamamoto, Y., 1987, "An Automated Guidance System of a Self-Controlled Vehicle." *IEEE Transactions on Industrial Electronics,* Vol. IE-34, No. 1, pp. 5-10.

Hoppen, P., Knieriemen, T., and Puttkamer, E., 1990, "Laser-Radar Based Mapping and Navigation for an Autonomous Mobile Robot." *Proceedings of IEEE International Conference on Robotics and Automation*, Cincinnati, OH, May 13-18, pp. 948-953.

Hurn, J., 1993, *GPS, A Guide to the Next Utility*, No. 16778, Trimble Navigation, Sunnyvale, CA, Nov..

Janet, J., Luo, R., Aras, C., and Kay, M., 1993, "Sonar Windows and Geometrically Represented Objects for Mobile Robot Self-Referencing." *Proceedings of the 1993 IEEE/RSJ International Conference on Intelligent Robotics and Systems*, Yokohama, Japan, July 26-30, pp. 1324-1331.

Jenkin, M., Milios, E., Jasiobedzki, P., Bains, N., and Tran, K., 1993, "Global Navigation for ARK." *Proceedings of the 1993 IEEE/RSJ International Conference on Intelligent Robotics and Systems*, Yokohama, Japan, July 26-30, pp. 2165 2171.

Jörg, K.W., 1994, "*Echtzeitfähige Multisensorintegration für autonome mobile Roboter.*" ISBN 3-411-16951-6, B.I. Wissenschaftsverlag, Mannheim, Leipzig, Wien, Zürich.

Jörg, K.W., 1995, "World Modeling for an Autonomous Mobile Robot Using Heterogenous Sensor Information." *Robotics and Autonomous Systems*, Vol. 14, pp. 159-170.

Jones, J.L. and Flynn, A., 1993, *Mobile Robots: Inspiration to Implementation*. ISBN 1-56881-011-3, A K Peters, Ltd. Wellesley, MA.

Kabuka, M and Arenas, A., 1987, "Position Verification of a Mobile Robot Using Standard Pattern." *IEEE Journal of Robotics and Automation*, Vol. RA-3, No. 6, pp. 505-516.

Kadonoff, M.B., 1986, "Navigation Techniques for the Denning Sentry."*MS86-757, RI/SME 2nd International Conference on Robotics Research*, Scottsdale, AZ, August.

Kak, A., Andress, K., Lopez-Abadia, and Carroll, M., 1990, "Hierarchical Evidence Accumulation in the PSEIKI System and Experiments in Model-driven Mobile Robot Navigation." in *Uncertainty in Artificial Intelligence*, Vol. 5, Elsevier Science Publishers B. V., North-Holland, pp. 353-369.

Kay, M. and Luo, R., 1993, "Global Vision for the Control of Free-Ranging AGV Systems." *Proceedings of IEEE International Conference on Robotics and Automation*, Atlanta, GA, May 10-15, pp. 14-19.

Kenny, T.W., Waltman, S.B., Reynolds, J.K., and Kaiser, W.J., 1991, "Micromachined Silicon Tunnel Sensor for Motion Detection." *Applied Physics Letters*, Vol. 58, No. 1, January.

Kerr, J.R., 1988, "Real Time Imaging Rangefinder for Autonomous Land Vehicles." *SPIE Vol. 1007, Mobile Robots III*, Cambridge, MA, November, pp. 349-356.

Kihara, M. and Okada, T., 1984, "A Satellite Selection Method and Accuracy for the Global Positioning System." *Navigation: Journal of the Institute of Navigation*, Vol. 31, No. 1, Spring., pp. 8-20.

Killough, S.M., Pin, F.G., 1992, "Design of an Omnidirectional Holonomic Wheeled Platform Prototype." *Proceedings of the IEEE Conference on Robotics and Automation*, Nice, France, May, pp. 84-90.

Kim, E.J., 1986, "Design of a Phased Sonar Array for a Mobile Robot." *Bachelor's Thesis, MIT*, Cambridge, MA, May.

King, S. and Weiman, C., 1990, "HelpMate Autonomous Mobile Robot Navigation System." *Proceedings of the 1990 SPIE Conference on Mobile Robots,* Boston, MA, Nov. 8-9, pp. 190-198.

Klarer, P.R., 1988, "Simple 2-D Navigation for Wheeled Vehicles." *Sandia Report SAND88-0540, Sandia National Laboratories*, Albuquerque, NM, April.

Kleeman, L., 1992, "Optimal Estimation of Position and Heading for Mobile Robots Using Ultrasonic Beacons and Dead-reckoning." *Proceedings of IEEE International Conference on Robotics and Automation*, Nice, France, May 12-14, pp. 2582-2587.

Kleeman, L. and Russell, R., 1993, "Thermal Path Following Robot Vehicle: Sensor Design and Motion Control." *Proceedings of the 1993 IEEE/RSJ International Conference on Intelligent Robotics and Systems*, Yokohama, Japan, July 26-30, pp. 1319-1323.

Koenigsburg, W.D., 1982, "Noncontact Distance Sensor Technology." *GTE Laboratories, Inc.*, 40 Sylvan Rd., Waltham, MA, 02254, March, pp. 519-531.

Komoriya, K. and Oyama, E., 1994, "Position Estimation of a Mobile Robot Using Optical Fiber Gyroscope (OFG)." *International Conference on Intelligent Robots and Systems (IROS '94)*. Munich, Germany, Sept. 12-16, pp. 143-149.

Koper, J.G., 1987, "A Three-Axis Ring Laser Gyroscope," *Sensors*, March, pp. 8-21.

Kortenkamp, D. and Weymouth, T., 1994, "Combining Sonar and Vision Sensing in the Construction and Use of Topological Maps for Mobile Robots." Submitted to *the IEEE Transactions on Robotics and Automation*.

Krotkov, E., 1991b, "Mobile Robot Localization Using a Single Image." *Proceedings of IEEE International Conference on Robotics and Automation*, Sacramento, CA, April 9-11, pp. 978-983.

Kuc, R., and Siegel, M.W., 1987, "A physically-based simulation model for acoustic sensor robot navigation." *IEEE Trans. Pattern Analysis and Machine Intelligence PAMI-9*, No. 6, pp. 766 -778.

Kwiatkowski, W. and Tumanski, S., 1986, "The Permalloy Magnetoresistive Sensors - Properties and Applications." *J. Phys. E: Sci. Instrum.*, Vol. 19, pp. 502-515.

La, W.H.T., Koogle, T.A., Jaffe, D.L., and Leifer, L.J., 1981, "Microcomputer-Controlled Omnidirectional Mechanism for Wheelchairs." *Proceedings, IEEE Frontiers of Engineering in Health Care*, CH1621-2/81/0000-0326.

Langer, D. and Thorpe, C., 1992, "Sonar Based Outdoor Vehicle Navigation and Collision Avoidance." *International Conference on Intelligent Robots and Systems, IROS '92*, Raleigh, NC, July.

Langley, R.B., 1991, "The Mathematics of GPS." *GPS World*, July/Aug., pp. 45-49.

Lapin, B., 1992, "Adaptive Position Estimation for an Automated Guided Vehicle." *Proceedings of the 1992 SPIE Conference on Mobile Robots,* Boston, MA, Nov. 18-20, pp. 82-94.

Larson, T.R. and Boltinghouse, S., 1988, "Robotic Navigation Within Complex Structures." *SPIE Vol. 1007, Mobile Robots III*, Cambridge, MA, Nov., pp. 339-348.

Larsson, U., Zell, C., Hyyppa, K., and Wernersson, A., 1994, "Navigating an Articulated Vehicle and Reversing with a Trailer." *Proceedings of IEEE International Conference on Robotics and Automation*, San Diego, CA, May 8-13, pp. 2398-2404.

Lefevre, H.C., 1992, "The Interferometric Fiber-Optic Gyroscope." in *Fiber Optic Sensors*, Udd, E., Editor, Vol. CR44, SPIE Optical Engineering Press, Bellingham, WA, Sept.

Lenz, J.E., 1990, "A Review of Magnetic Sensors." *Proceedings of the IEEE*, Vol. 78, No. 6, June.

Leonard, J. and Durrant-Whyte, H.F., 1990, "Application of Multi-Target Tracking to Sonar-Based Mobile Robot Navigation." *International Conference on Decision and Control*.

Leonard, J. and Durrant-Whyte, H. F., 1991, "Mobile Robot Localization by Tracking Geometric Beacons." *IEEE Transactions on Robotics and Automation*, Vol. 7, No. 3, pp. 376-382.

Lewis, R.A. and Johnson, A.R., 1977, "A Scanning Laser Rangefinder for a Robotic Vehicle." *5th International Joint Conference on Artificial Intelligence*, pp. 762-768.

MacLeod, E. and Chiarella, M., 1993, "Navigation and Control Breakthrough for Automated Mobility." *Proceedings of the 1993 SPIE Conference on Mobile Robots,* Boston, MA, Sept. 9-10, pp. 57-68.

Maddox, J., 1994, "Smart Navigation Sensors for Automatic Guided Vehicles." *Sensors*, April, pp. 48-50.

Maenaka, K., Ohgusu, T., Ishida, M., and Nakamura, T., 1987, "Novel Vertical Hall Cells in Standard Bipolar Technology, *Electronic Letters*, Vol. 23, pp. 1104-1105.

Maenaka, K., Tsukahara, M., and Nakamura, T., 1990, "Monolithic Silicon Magnetic Compass." *Sensors and Actuators*, pp. 747-750.

Magee, M. and Aggarwal, J., 1984, "Determining the Position of a Robot Using a Single Calibrated Object." *Proceedings of IEEE International Conference on Robotics and Automation*, Atlanta, GA, March 13-15, pp. 140-149.

Mahajan, A., 1992, "A Navigation System for Guidance and Control of Autonomous Vehicles Based on an Ultrasonic 3-D Location System." *Master's Thesis, Mechanical Engineering Department, Tulane University*, July.

Manolis, S., 1993, "Resolvers vs. Rotary Encoders For Motor Commutation and Position Feedback." *Sensors*, March, pp. 29-32.

Martin, G.J., 1986, "Gyroscopes May Cease Spinning." *IEEE Spectrum*, February, pp. 48-53.

Komoriya, K. and Oyama, E., 1994, "Position Estimation of a Mobile Robot Using Optical Fiber Gyroscope (OFG)." *International Conference on Intelligent Robots and Systems (IROS '94)*. Munich, Germany, Sept. 12-16, pp. 143-149.

Koper, J.G., 1987, "A Three-Axis Ring Laser Gyroscope," *Sensors*, March, pp. 8-21.

Kortenkamp, D. and Weymouth, T., 1994, "Combining Sonar and Vision Sensing in the Construction and Use of Topological Maps for Mobile Robots." Submitted to *the IEEE Transactions on Robotics and Automation.*

Krotkov, E., 1991b, "Mobile Robot Localization Using a Single Image." *Proceedings of IEEE International Conference on Robotics and Automation*, Sacramento, CA, April 9-11, pp. 978-983.

Kuc, R., and Siegel, M.W., 1987, "A physically-based simulation model for acoustic sensor robot navigation." *IEEE Trans. Pattern Analysis and Machine Intelligence PAMI-9*, No. 6, pp. 766 -778.

Kwiatkowski, W. and Tumanski, S., 1986, "The Permalloy Magnetoresistive Sensors - Properties and Applications." *J. Phys. E: Sci. Instrum.*, Vol. 19, pp. 502-515.

La, W.H.T., Koogle, T.A., Jaffe, D.L., and Leifer, L.J., 1981, "Microcomputer-Controlled Omnidirectional Mechanism for Wheelchairs." *Proceedings, IEEE Frontiers of Engineering in Health Care*, CH1621-2/81/0000-0326.

Langer, D. and Thorpe, C., 1992, "Sonar Based Outdoor Vehicle Navigation and Collision Avoidance." *International Conference on Intelligent Robots and Systems, IROS '92*, Raleigh, NC, July.

Langley, R.B., 1991, "The Mathematics of GPS." *GPS World*, July/Aug., pp. 45-49.

Lapin, B., 1992, "Adaptive Position Estimation for an Automated Guided Vehicle." *Proceedings of the 1992 SPIE Conference on Mobile Robots,* Boston, MA, Nov. 18-20, pp. 82-94.

Larson, T.R. and Boltinghouse, S., 1988, "Robotic Navigation Within Complex Structures." *SPIE Vol. 1007, Mobile Robots III*, Cambridge, MA, Nov., pp. 339-348.

Larsson, U., Zell, C., Hyyppa, K., and Wernersson, A., 1994, "Navigating an Articulated Vehicle and Reversing with a Trailer." *Proceedings of IEEE International Conference on Robotics and Automation*, San Diego, CA, May 8-13, pp. 2398-2404.

Lefevre, H.C., 1992, "The Interferometric Fiber-Optic Gyroscope." in *Fiber Optic Sensors*, Udd, E., Editor, Vol. CR44, SPIE Optical Engineering Press, Bellingham, WA, Sept.

Lenz, J.E., 1990, "A Review of Magnetic Sensors." *Proceedings of the IEEE*, Vol. 78, No. 6, June.

Leonard, J. and Durrant-Whyte, H.F., 1990, "Application of Multi-Target Tracking to Sonar-Based Mobile Robot Navigation." *International Conference on Decision and Control.*

Leonard, J. and Durrant-Whyte, H. F., 1991, "Mobile Robot Localization by Tracking Geometric Beacons." *IEEE Transactions on Robotics and Automation*, Vol. 7, No. 3, pp. 376-382.

Lewis, R.A. and Johnson, A.R., 1977, "A Scanning Laser Rangefinder for a Robotic Vehicle." *5th International Joint Conference on Artificial Intelligence*, pp. 762-768.

MacLeod, E. and Chiarella, M., 1993, "Navigation and Control Breakthrough for Automated Mobility." *Proceedings of the 1993 SPIE Conference on Mobile Robots,* Boston, MA, Sept. 9-10, pp. 57-68.

Maddox, J., 1994, "Smart Navigation Sensors for Automatic Guided Vehicles." *Sensors*, April, pp. 48-50.

Maenaka, K., Ohgusu, T., Ishida, M., and Nakamura, T., 1987, "Novel Vertical Hall Cells in Standard Bipolar Technology, *Electronic Letters*, Vol. 23, pp. 1104-1105.

Maenaka, K., Tsukahara, M., and Nakamura, T., 1990, "Monolithic Silicon Magnetic Compass." *Sensors and Actuators*, pp. 747-750.

Magee, M. and Aggarwal, J., 1984, "Determining the Position of a Robot Using a Single Calibrated Object." *Proceedings of IEEE International Conference on Robotics and Automation*, Atlanta, GA, March 13-15, pp. 140-149.

Mahajan, A., 1992, "A Navigation System for Guidance and Control of Autonomous Vehicles Based on an Ultrasonic 3-D Location System." *Master's Thesis, Mechanical Engineering Department, Tulane University*, July.

Manolis, S., 1993, "Resolvers vs. Rotary Encoders For Motor Commutation and Position Feedback." *Sensors*, March, pp. 29-32.

Martin, G.J., 1986, "Gyroscopes May Cease Spinning." *IEEE Spectrum*, February, pp. 48-53.

Mataric, M., 1990, "Environment Learning Using a Distributed Representation." *Proceedings of IEEE International Conference on Robotics and Automation*, Cincinnati, OH, May 13-18, pp. 402-406.

Matsuda, T. and Yoshikawa, E., 1989, "Z-shaped Position Correcting Landmark for AGVs." *Proceedings of the 28th SICE Annual Conference*, July 25-27, pp. 425-426.

Matsuda T. et al., 1989, "Method of Guiding an Unmanned Vehicle." U.S. Patent #4,866,617. Issued Sep.12.

McGillem, C. and Rappaport, T., 1988, "Infra-red Location System for Navigation of Autonomous Vehicles." *Proceedings of IEEE International Conference on Robotics and Automation*, Philadelphia, PA, April 24-29, pp. 1236-1238.

McPherson, J.A., 1991, "Engineering and Design Applications of Differential Global Positioning Systems (DGPS) for Hydrographic Survey and Dredge Positioning." *Engineering Technical Letter No 1110-1-150, US Army Corps of Engineers*, Washington, DC, July 1.

Menegozzi, L.N., Lamb, W.E., 1973, "Theory of a Ring Laser." *Physical Review A*, Vol. 1, No. 4, October, pp. 2103-2125.

Mesaki, Y. and Masuda, I., 1992, "A New Mobile Robot Guidance System Using Optical Reflectors." *Proceedings of the 1992 IEEE/RSJ International Conference on Intelligent Robots and Systems*, Raleigh, NC, July 7-10, pp. 628-635.

Miller, G.L. and Wagner, E.R., 1987, "An Optical Rangefinder for Autonomous Robot Cart Navigation." *Proceedings of the Advances in Intelligent Robotic Systems: SPIE Mobile Robots II.*

Moravec, H.P. and Elfes, A., 1985, "High Resolution Maps from Wide Angle Sonar." *Proceedings of the IEEE Conference on Robotics and Automation*, Washington, D.C., pp. 116-121.

Moravec, H.P., 1988, "Sensor Fusion in Certainty Grids for Mobile Robots." *AI Magazine*, Summer, pp. 61-74.

Motazed, B., 1993, "Measure of the Accuracy of Navigational Sensors for Autonomous Path Tracking." *Proceedings, SPIE Vol. 2058, Mobile Robots VIII*, pp. 240-249.

Murray, C., 1991, "AGVs Go Wireless." *Design News*, June, pp. 27-28.

Nickson, P., 1985, "Solid-State Tachometry." *Sensors*, April, pp. 23-26.

Nishide, K., Hanawa, M., and Kondo, T., 1986, "Automatic Position Findings of Vehicle by Means of Laser." *Proceedings of IEEE International Conference on Robotics and Automation*, San Francisco, CA, Apr. 7-10, pp. 1343-1348.

Nitzan, D. et al. 1977, "The Measurement and Use of Registered Reflectance and Range Data in Scene Analysis." *Proceedings of IEEE*, Vol. 65, No. 2, Feb., pp. 206-220.

Nolan, D.A., Blaszyk, P.E., and Udd, E., 1991, "Optical Fibers." *Fiber Optic Sensors: An Introduction for Engineers and Scientists*, E. Udd, Ed., John Wiley and Sons, Inc., New York, pp. 9-26.

Parish, D. and Grabbe, R., 1993, "Robust Exterior Autonomous Navigation." *Proceedings of the 1993 SPIE Conference on Mobile Robots,* Boston, MA, Sept. 9-10, pp. 280-291.

Patterson, M.R., Reidy, J.J., and Rudolph, R.C., 1984, "Guidance and Actuation Systems for an Adaptive-Suspension Vehicle." *Final Technical Report, Battelle Columbus Division*, OH, AD#A139111, March 20.

Pessen, D.W., 1989, "*Industrial Automation.*" ISBN 0-471-60071-7, John Wiley and Sons, Inc.

Petersen, A., 1989, "Magnetoresistive Sensors for Navigation." *Proceedings, 7th International Conference on Automotive Electronics*, London, England, Oct, pp. 87-92.

Pin, F.G. and Killough, M., 1994, "A New Family of Omnidirectional and Holonomic Wheeled Platforms for Mobile Robots." *IEEE Transactions on Robotics and Automation*, Vol. 10, No. 4, Aug., pp. 480-489.

Pin, F.G. et al., 1989, "Autonomous Mobile Robot Research Using the HERMIES-III Robot." *IROS International Conference on Intelligent Robot and Systems*, Tsukuba, Japan, Sept.

Pin, F.G. and Watanabe, Y., 1993, "Using Fuzzy Behaviors for the Outdoor Navigation of a Car with Low-Resolution Sensors." *IEEE International Conference on Robotics and Automation*, Atlanta, Georgia, May 2-7, pp. 548-553.

Pletta, J.B., Amai, W.A., Klarer, P., Frank, D., Carlson, J., and Byrne, R., 1992, "The Remote Security Station (RSS) Final Report." *Sandia Report SAND92-1947 for DOE under Contract DE-AC04-76DP00789, Sandia National Laboratories*, Albuquerque, NM, Oct.

Premi, K.S. and Besant, C.B., 1983, "A Review of Various Vehicle Guidance Techiques That Can be Used by Mobile Robots or AGVS." *2nd International Conference on Automated Guided Vehicle Systems*, Stuttgart, Germany, June.

Primdahl, F., 1970, "The Fluxgate Mechanism, Part I: The Gating Curves of Parallel and Orthogonal Fluxgates." *IEEE Transactions on Magnetics*, Vol. MAG-6, No. 2, June.

Primdahl, F., 1979, "The Fluxgate Magnetometer." *J. Phys. E: Sci. Instrum.*, Vol. 12, pp. 241-253.

Purkey, M., 1994, "On Target." *Golf Magazine*, May, pp. 120-121.

Raschke, U. and Borenstein, J., 1990, "A Comparison of Grid-type Map-building Techniques by Index of Performance." *Proceedings of IEEE International Conference on Robotics and Automation*, Cincinnati, CA, May 13-18, pp. 1828-1832.

Reister, D.B., 1991, "A New Wheel Control System for the Omnidirectional HERMIES-III Robot." *Proceedings of the IEEE Conference on Robotics and Automation*, Sacramento, California, April 7-12, pp. 2322-2327.

Reister, D.B. et al., 1991, "DEMO 89 — The Initial Experiment With the HERMIES-III Robot." *Proceedings of the 1991 IEEE Conference on Robotics and Automation* Sacramento, California, April, pp. 2562-2567.

Reister, D.B. and Unseren, M.A., 1992, "Position and Force Control of a Vehicle with Two or More Steerable Drive Wheels." *Internal Report ORNL/TM-12193, Oak Ridge National Laboratories*.

Reister, D.B. and Unseren, M.A., 1993, "Position and Constraint Force Control of a Vehicle with Two or More Steerable Drive Wheels." *IEEE Transactions on Robotics and Automation.*Vol. 9, No. 6, December, pp. 723-731.

Rencken, W.D., 1993, "Concurrent Localization and Map Building for Mobile Robots Using Ultrasonic Sensors." *Proceedings of the 1993 IEEE/RSJ International Conference on Intelligent Robotics and Systems*, Yokohama, Japan, July 26-30, pp. 2192-2197.

Rencken, W.D., 1994, "Autonomous Sonar Navigation in Indoor, Unknown, and Unstructured Environments."*1994 International Conference on Intelligent Robots and Systems (IROS '94)*. Munich, Germany, Sept. 12-16, pp. 127-134.

Reunert, M.K., 1993, "Fiber Optic Gyroscopes: Principles and Applications." *Sensors*, August, pp. 37-38.

Russell, R.A., Thiel, D., and Mackay-Sim, A., 1994, "Sensing Odor Trails for Mobile Robot Navigation." *Proceedings of IEEE International Conference on Robotics and Automation*, San Diego, CA, May 8-13, pp. 2672-2677.

Russell, R.A. 1993, "Mobile Robot Guidance Using a Short-lived Heat Trail." *Robotica,* Vol 11, Part 5, pp. 427-431.

Russell, R.A., 1995a, "A Practical Demonstration of the Application of Olfactory Sensing to Robot Navigation." *Proceedings of the International Advanced Robotics Programme (IARP)*, Sydney, Australia, May 18-19, pp. 35-43.

Russell, R.A., 1995b, "Laying and Sensing Odor Markings as a Strategy for Assisting Mobile Robot Navigation Tasks." *IEEE Robotics and Automation Magazine*, Vol. 2, No. 3, Sept., pp. 3-9.

Sabatini, A. and Benedetto, O., 1994, "Towards a Robust Methodology for Mobile Robot Localization Using Sonar." *Proceedings of IEEE International Conference on Robotics and Automation*, San Diego, CA, May 8-13, pp. 3142-3147.

Sagnac, G.M., 1913, "L'ether lumineux demontre par l'effet du vent relatif d'ether dans un interferometre en rotation uniforme." *C.R. Academy of Science*, 95, pp. 708-710.

Sammarco, J.J., 1994, "A Navigational System for Continuous Mining Machines." *Sensors*, Jan., pp. 11-17.

Sammarco, J.J., 1990, "Mining Machine Orientation Control Based on Inertial, Gravitational, and Magnetic Sensors." *Report of Investigations 9326*, US Bureau of Mines, Pittsburgh, PA.

Sanders, G.A., 1992, "Critical Review of Resonator Fiber Optic Gyroscope Technology." in *Fiber Optic Sensors*, Udd, E., Ed., Vol. CR44, SPIE Optical Engineering Press, Bellingham, WA, Sept.

Schaffer, G., Gonzalez, J., and Stentz, A., 1992, "Comparison of Two Range-based Pose Estimators for a Mobile Robot." *Proceedings of the 1992 SPIE Conference on Mobile Robots,* Boston, MA, Nov. 18-20, pp. 661-667.

Schiele, B. and Crowley, J., 1994, "A Comparison of Position Estimation Techniques Using Occupancy Grids." *Proceedings of IEEE International Conference on Robotics and Automation*, San Diego, CA, May 8-13, pp. 1628-1634.

Schiele, B. and Crowley, J., 1994, "A Comparison of Position Estimation Techniques Using Occupancy Grids." *Robotics and Autonomous Systems*, Vol. 12, pp. 163-171.

Schultz, W., 1993, "Traffic and Vehicle Control Using Microwave Sensors." *Sensors*, October, pp. 34-42.

Schulz-DuBois, E.O., 1966, "Alternative Interpretation of Rotation Rate Sensing by Ring Laser." *IEEE Journal of Quantum Electronics*, Vol. QE-2, No. 8, Aug., pp. 299-305.

Shoval, S., Benchetrit, U., and Lenz, E., 1995, "Control and Positioning of an AGV for Material Handling in an Industrial Environment." *Proceedings of the 27th CIRP International Seminar on Manufacturing Systems*, Ann Arbor, MI, May 21-23, pp. 473-479.

Siuru, B., 1994, "The Smart Vehicles Are Here." *Popular Electronics*, Vol. 11, No. 1, Jan., pp. 41-45.

Stokes, K.W., 1989, "Remote Control Target Vehicles for Operational Testing." Association for Unmanned Vehicles Symposium, Washington, DC, July.

Stuart, W.F., 1972, "Earth's Field Magnetometry, *Reports on Progress in Physics*, J.M. Zinman, Ed., Vol. 35, Part 2, pp. 803-881.

Stuck, E. R., Manz, A., Green, D. A., and Elgazzar, S., 1994, "Map Updating and Path Planning for Real-Time Mobile Robot Navigation."*1994 International Conference on Intelligent Robots and Systems (IROS '94)*. Munich, Germany, Sept. 12-16, pp. 753-760.

Sugiyama, H., 1993, "A Method for an Autonomous Mobile Robot to Recognize its Position in the Global Coordinate System when Building a Map." *Proceedings of the 1993 IEEE/RSJ International Conference on Intelligent Robotics and Systems*, Yokohama, Japan, July 26-30, pp. 2186-2191.

Tai, S., Kojima, K., Noda, S., Kyuma, K., Hamanaka, K., and Nakayama, T., 1986, "All-Fibre Gyroscope Using Depolarized Superluminescent Diode." *Electronic Letters*, Vol. 22, p. 546.

Talluri, R., and Aggarwal, J., 1993, "Position Estimation Techniques for an Autonomous Mobile Robot - a Review." in *Handbook of Pattern Recognition and Computer Vision*, World Scientific: Singapore, Chapter 4.4, pp. 769-801.

Takeda, T., Kato, A., Suzuki, T., and Hosoi, M., 1986, "Automated Vehicle Guidance Using Spotmark." *Proceedings of IEEE International Conference on Robotics and Automation*, San Francisco, CA, Apr. 7-10, pp. 1349-1353.

Taylor, C., 1991, "Building Representations for the Environment of a Mobile Robot from Image Data." *Proceedings of the 1991 SPIE Conference on Mobile Robots,* Boston, MA, Nov. 14-15, pp. 331-339.

Tonouchi, Y., Tsubouchi, T., and Arimoto, S., 1994, "Fusion of Dead-reckoning Positions With a Workspace Model for a Mobile Robot by Bayesian Inference." *International Conference on Intelligent Robots and Systems (IROS '94)*. Munich, Germany, Sept. 12-16, pp. 1347-1354.

Tsumura, T. and Hashimoto, M., 1986, "Positioning and Guidance of Ground Vehicle by Use of Laser and Corner Cube." *Proceedings of IEEE International Conference on Robotics and Automation*, San Francisco, CA, Apr. 7-10, pp. 1335-1342.

Tsumura, T., 1986, "Survey of Automated Guided Vehicle in Japanese Factory." *Proceedings of IEEE International Conference on Robotics and Automation*, San Francisco, CA, Apr. 7-10, pp. 1329-1334.

Tsumura, T., Fujiwara, N., Shirakawa, T., and Hashimoto, M.,1981, "An Experimental System for Automatic Guidance of Roboted Vehicle Following the Route Stored in Memory." *Proc. of the 11th Int. Symp. on Industrial Robots*, Tokyo, Japan, pp. 18-193.

Tsumura, T., Hashimoto, M., and Fujiwara, N., 1988, "A Vehicle Position and Heading Measurement System Using Corner Cube and Laser Beam." *Proceedings of IEEE International Conference on Robotics and Automation*, Philadelphia, PA, Apr. 24-29, pp. 47-53.

Turpin, D.R., 1986, "Inertial Guidance: Is It a Viable Guidance System for AGVs?" *4th International Conference on AGVs (AGVS4)*, June, pp. 301-320.

Udd, E., 1985, "Fiberoptic vs. Ring Laser Gyros: An Assessment of the Technology." in *Laser Focus/Electro Optics*, Dec.

Udd, E., 1991, "Fiberoptic Sensors Based on the Sagnac Interferometer and Passive Ring Resonator." in *Fiber Optic Sensors: An Introduction for Engineers and Scientists*, E. Udd, Ed., John Wiley and Sons, Inc., New York, pp. 233-269.

Vaganay, J., Aldon, M.J., and Fournier, A., 1993a, "Mobile Robot Attitude Estimation by Fusion of Inertial Data." *Proceedings of IEEE International Conference on Robotics and Automation*, Atlanta, GA, May 10-15, pp. 277-282.

Vaganay, J., Aldon, M.J., and Fournier, A., 1993b, "Mobile Robot Localization by Fusing Odometric and Inertial Measurements." 5th Topical Meeting on Robotics and Remote Systems, Knoxville, TN, Vol. 1, Apr., pp. 503-510.

Vestli, S.J., Tschichold-Gürman, N., Adams, M., and Sulzberger, S., 1993, "Amplitude Modulated Optical Range Data Analysis in Mobile Robotics." *Proceedings of the 1993 IEEE International Conference on Robotics and Automation*, Atlanta, GA, May 2-7, pp 3.243 - 3.248.

Vuylsteke, P., Price, C.B., and Oosterlinck, A., 1990, "Image Sensors for Real-Time 3D Acquisition, Part 1." *Traditional and Non-Traditional Robotic Sensors*, T.C. Henderson, Ed., NATO ASI Series, Vol. F63, Springer-Verlag, pp. 187-210.

Wax, S.I. and Chodorow, M., 1972, "Phase Modulation of a Ring-Laser Gyro - Part II: Experimental Results," *IEEE Journal of Quantum Electronics*, March, pp. 352-361.

Weiß, G., Wetzler, C., and Puttkamer, E., 1994, "Keeping Track of Position and Orientation of Moving Indoor Systems by Correlation of Range-Finder Scans." *1994 International Conference on Intelligent Robots and Systems (IROS'94)*, Munich, Germany, Sept. 12-16, pp. 595-601.

Wienkop, U., Lawitzky, G., and Feiten, W., 1994, "Intelligent Low-cost Mobility." *1994 International Conference on Intelligent Robots and Systems (IROS '94)*. Munich, Germany, Sept. 12-16, pp. 1708-1715.

Wiley, C.M., 1964, "Navy Tries Solid-State Compass." *Electronics*, Feb. 14, pp. 57-58.

Wilkinson, J.R., 1987, "Ring Lasers." *Progress in Quantum Electronics*, edited by Moss, T.S., Stenholm, S., Firth, W.J., Phillips, W.D., and Kaiser, W., Vol. 11, No. 1, Pergamon Press, Oxford.

Woll, J.D., 1993, "A Review of the Eaton VORAD Vehicle Collision Warning System." Reprinted from *International Truck and Bus Meeting and Exposition*, Detroit, MI, SAE Technical Paper Series 933063, ISSN 0148-7191 Nov., pp. 1-4.

Wong, A. and Gan, S., "Vision Directed Path Planning, Navigation, and Control for An Autonomous Mobile Robot." *Proceedings of the 1992 SPIE Conference on Mobile Robots*, Boston, MA, Nov. 18-20, pp. 352-360.

Woodbury, N., Brubacher, M., and Woodbury, J.R., 1993, "Noninvasive Tank Gauging with Frequency-Modulated Laser Ranging." *Sensors*, Sept., pp. 27-31.

Wormley, S., 1994, "A Little GPS Background." Internet message, swormley@ thrl.cnde.iastate.edu to Newsgroup sci.geo.satellite-nav, March 15.

Wun-Fogle, M. and Savage, H.T., 1989, "A Tunneling-tip Magnetometer." *Sensors and Actuators*, Vol. 20, pp. 199-205.

Cited Product Information from Commercial Companies

ACUITY - Acuity Research, POC: Bob Clark, 20863 Stevens Creek Blvd, Cupertino, CA 95014-2115, 415 369-6782.

ADL - Applied Design Laboratories, P. O. Box 2405, Grass Valley, CA 95945, 916-272-8206

AECL - Atomic Energy of Canada Ltd., Sheridan Research Park, 2251 Speakman Drive, Mississauga, Ontario, L5K 1B2, Canada. POC Narindar Baines, 905-823-9060

ANDREW Andrew Corporation, 10500 W. 153rd Street, Orland Park, IL 60462. 708-349-5294 or 708-349-3300.

BENTHOS - Benthos, Inc., 49 Edgerton Drive, North Falmouth, MA 02556-2826,508-563-1000.

CATERPILLAR - Caterpillar Industrial, Inc., Product Literature, SGV-1106/91, Caterpillar Self Guided Vehicle Systems, 5960 Heisley Rd, Painesville, OH 44077, 216-357-2246.

CONTROL - Control Engineering Company, 8212 Harbor Spring Road, Harbor Spring, MI 49740, 616-347-3931.

CYBERMOTION - Cybermotion, Inc., 115 Sheraton Drive, Salem, VA 24153, 703-562-7626.

CYBERWORKS - Cyberworks, Inc., "Camera Vision Robot Position and Slippage Control System." Product Literature, 31 Ontario Street, Orillia, Ontario, L3V 6H1 Canada, 705-325-6110.

DBIR - Denning Branch International Robotics, 1401 Ridge Avenue, Pittsburgh PA 15233, 412-322-4412.

DINSMORE - Dinsmore Instrument Company, Product Literature, 1490 and 1525 Magnetic Sensors, Flint, MI, 313-744-1330.

ERIM - Environmental Research Institute of Michigan, Box 8618, Ann Arbor, MI 48107, 313-994-1200.

EATON - Eaton-Kenway, Inc., 515 East 100 South, 515 E 100 S, Salt Lake City, UT 84102, 801-530-4688.

ESP - ESP Technologies, Inc., "ORS-1 Optical Ranging System." Product Literature, ESP Technologies, Inc., 21 Le Parc Drive, Lawrenceville, NJ 08648, 609-275-0356.

FUTABA - Futaba Corporation of America, 4 Studebaker, Irvine, CA 92718, 714-455-9888.

GEC - GEC Avionics, Kent, U.K.

GPS - GPS Report, 1992, Phillips Business Information, Potomac, MD, Nov.

GREYHOUND - 1994, "Accident Rate Keeps Falling." *Greyhound Connections*, Vol. 4, No. 2, March/April.

GYRATION - Gyration, Inc., 12930 Saratoga Ave., Building C, Saratoga, CA 95070-4600, 408-255-3016.

HITACHI - Hitachi Cable America, Inc., New York Office, 50 Main Street, 12th floor, White Plains, NY 10606, 914-993-0990.

HP - Hewlett Packard Components, "Optoelectronics Designer's Catalog, 1991-1992, 19310 Pruneridge Ave., Cupertino, CA, 800-752-9000.

HTI - Harris Technologies, Inc., PO Box 6, Clifton, VA 22024, 703-266-0904.

ILC - ILC Data Device Corporation, 1982, "Synchro Conversion Handbook," Bohemia, NY.

ISI - Intelligent Solutions, Inc., EZNav Position Sensor, One Endicott Avenue, Marblehead, MA 01945, 617-646-4362.

ISR - IS Robotics, Inc., RR-1/BS-1 System for Communications and Positioning. Preliminary Data Sheet." IS Robotics, Twin City Office Center, Suite 6, 22 McGrath Highway, Somerville, MA 02143, 617-629-0055.

KAMAN - Kaman Sciences Corporation, "Threat Array Control and Tracking Information Center." Product Literature, PM1691, Colorado Springs, CO, 719-599-1285.

KVH - KVH Industries, C100 Compass Engine Product Literature, 110 Enterprise Center, Middletown, RI 02840, 401-847-3327.

MASSA - Massa Products Corporation, "E-201B & E-220B Ultrasonic Ranging Module Subsystems Product Selection Guide." Product Literature 891201-10M, Hingham, MA 02043, 617-749-4800.

MICRO-TRAK - Micro-Trak Systems, Inc., "Trak-Star Ultrasonic Speed Sensor." Product Information. P.O. Box 3699, Mankato, MN 56002, 507-257-3600.

MTI - MTI Research, Inc., "Computerized Opto-electronic Navigation and Control (CONAC™)" and "What You Can Expect From CONAC™ Products." Product literature. 313 Littleton Road, Chelmsford, MA 01824., 508-250-4949.

MOTOROLA - Mini-Ranger Falcon, Product Literature, Motoroloa Government and Systems Technology Group, 8220 E. Roosevelt Road, PO Box 9040, Scottsdale, AZ 85252, 602-441-7685.

MURATA - Murata Erie North America, 2200 Lake Park Drive, Smyrna, GA 30080, 800-831-9172.

NAMCO - Namco Controls, 7567 Tyler Blvd. Mentor, OH 44060, 800-626-8324.

NASA - 1977, "Fast, Accurate Rangefinder." *NASA Tech Brief*, NPO-13460.

NIKE - NIKE, Dept. of Fitness Electronics, 9000 South-West Nimbus, Beaverton, Oregon 97005, 503-644-9000.

POLAROID - 1981, "Polaroid Ultrasonic Ranging System User's Manual." Publication No. P1834B, Polaroid Corporation, 784 Memorial Drive, Cambridge, MA 02139, 617-386-3964.

POLAROID - 1987, "Technical Specifications for Polaroid Electrostatic Transducer." 7000-Series Product Specification ITP-64, Polaroid Corporation, 784 Memorial Drive, Cambridge, MA 02139, 617-386-3964.

POLAROID - 1990, "6500-Series Sonar Ranging Module." Product Specifications PID 615077, Polaroid Corporation, 784 Memorial Drive, Cambridge, MA 02139, 617-386-3964.

POLAROID - 1991, "Ultrasonic Ranging System." Product Literature, Polaroid Corporation, 784 Memorial Drive, Cambridge, MA 02139, 617-386-3964.

POLAROID - 1993, "Polaroid Ultrasonic Ranging Developer's Kit." Publication No. PXW6431 6/93, Polaroid Corporation, 784 Memorial Drive, Cambridge, MA 02139, 617-386-3964.

REMOTEC - Remotec, 114 Union Valley Road, Oak Ridge, TN 37830, 615-483-0228

RIEGL - 1994, "Laser Distance, Level, and Speed Sensor LD90-3." Product Data Sheet 3/94, RIEGL Laser Measurement Systems, RIEGL USA, 4419 Parkbreeze Court, Orlando, FL 32808, 407-294-2799.

SFS - Safety First Systems, Ltd., POC: Allen Hersh, Safety First Systems Inc, 550 Stewart Ave, Garden City, NY 11530-4700, 516-681-3653.

SEO - 1991a, Schwartz Electro-Optics, Inc, "Scanning Laser Rangefinder." Product Literature, 3404 N. Orange Blossom Trail, Orlando, FL 32804, 407-297-1794.

SEO - 1991b, Schwartz Electro-Optics, Inc, Process Report for US Army Contract DAAJ02-91-C-0026, 3404 N. Orange Blossom Trail, Orlando, FL 32804, 407-297-1794.

SEO - 1995a, Schwartz Electro-Optics, Inc, "LRF-200 Laser Rangefinder Series." Product Literature, Schwartz Electro-Optics, Inc., 3404 N. Orange Blossom Trail, Orlando, FL 32804, 407-297-1794.

SEO - 1995b, Schwartz Electro-Optics, Inc., "SHIELD Scanning Helicopter Interference Envelope Laser Detector" Product Literature, 3404 N. Orange Blossom Trail, Orlando, FL 32804, 407-297-1794.

SIMAN - Siman Sensors & Intelligent Machines Ltd., MTI-Misgav, D.N. Misgav 20179, Israel, +972-4-906888.

SPERRY - Sperry Marine Inc., Seminole Trail, Charlottesville, VA, 22901, POC: Peter Arnold, Head of Engineering, ext. 2213, 804-974-2000.

TOWER - Tower Hobbies, Mail Order Catalog, P.O. Box 9078, Champaign, IL 61826-9078, 217-398-1100.

TRC - Transitions Research Corp., "TRC Light Ranger," Product Literature, Danbury, CT 06810, 203-798-8988.

TRC - Transitions Research Corp., "Beacon Navigation System," Product Literature, Shelter Rock Lane, Danbury, CT 06810, 203-798-8988.

UNIQUE -Unique Mobility, Inc., Product Literature, 425 Corporate Circle, Golden, CO 80401, 303-278-2002.

VORAD-1 - VORAD Safety Systems, Inc., "The VORAD Vehicle Detection and Driver Alert System." Product Literature, 10802 Willow Ct, San Diego, CA 92127, 619-674-1450.

VORAD-2 - Eaton VORAD Technologies, L.L.C., Eaton Center, Cleveland, OH 44114-2584, 216-523-5000

WATSON - Watson Industries, Inc., Melby Rd., Eau Claire, WI 54703, 715-839-0628.

ZEMCO - Zemco Group, Inc., 3401 Crow Canyon Road, Suite 201, San Ramon, CA 94583, 415-866-7266.

Relevant Research Articles on Mobile Robot Positioning, for Further Reading

Adams, M., 1993, "Amplitude Modulated Optical Range Data Analysis in Mobile Robots." *Proceedings of IEEE International Conference on Robotics and Automation*, Atlanta, GA, May 10-15, pp. 8-13.

Bhanu, B., Roberts, B., and Ming, J., 1990, "Inertial Navigation Sensor Integrated Motion Analysis for Obstacle Detection." *Proceedings of IEEE International Conference on Robotics and Automation*, Cincinnati, OH, May 13-18, pp. 954-959.

Blais, F., Rioux, M., and Domey, J., 1991, "Optical Range Image Acquisition for the Navigation of a Mobile Robot." *Proceedings of IEEE International Conference on Robotics and Automation*, Sacramento, CA, Apr. 9-11, pp. 2574-2580.

Bourbakis, N., 1988, "Design of an Autonomous Navigation System." *IEEE Control Systems Magazine*, Oct., pp. 25-28.

Chen, Q., Asada, M., and Tsuji, S., 1988, "A New 2-D World Representation System for Mobile Robots." *Proceedings of IEEE International Conference on Robotics and Automation*, Philadelphia, Apr. 24-29, pp. 604-606.

Crowley, J., 1986, "Navigation for an Intelligent Mobile Robot." *IEEE Journal of Robotics and Automation*, Vol. RA-1, No. 1, pp. 31-41.

Curran, A. and Kyriakopoulos, K., 1993, "Sensor-based Self-localization for Wheeled Mobile Robots." *Proceedings of IEEE International Conference on Robotics and Automation*, Atlanta, GA, May 10-15, pp. 8-13.

Drake, K., McVey, E., and Inigo, R., 1985, "Sensing Error for a Mobile Robot Using Line Navigation." *IEEE Transactions on Pattern Analysis and Machine Intelligence*, Vol. PAMI-7, No. 4, pp. 485-490.

Drake, K., McVey, E., and Inigo, R., 1987, "Experimental Position and Ranging Results for a Mobile Robot." *IEEE Journal of Robotics and Automation*, Vol. RA-3, No. 1, pp. 31-42.

Fennema, C., Hanson, A., Riseman, E., Beveridge, J., and Kumar, R., 1990, "Model-Directed Mobile Robot Navigation." *IEEE Transactions on Systems, Man, and Cybernetics*, Vol. 20, No. 6, pp. 1352-1369.

Harmon, S.Y., 1987, "The Ground Surveillance Robot (GSR): An Autonomous Vehicle Designed to Transit Unknown Terrain." *IEEE Journal of Robotics and Automation*, Vol. RA-3, No. 3, pp. 266-279.

Holcombe, W., Dickerson, S., Larsen, J., and Bohlander, R., 1988, "Advances in Guidance Systems for Industrial Automated Guided Vehicle." *Proceedings of the 1988 SPIE Conference on Mobile Robots,* Cambridge, MA, Nov. 10-11, pp. 288-297.

Huang, Y., Cao, Z., Oh, S., Katten, E., and Hall, E., 1986, "Automatic Operation for a Robot Lawn Mower." *Proceedings of the 1986 SPIE Conference on Mobile Robots,* Cambridge, MA, Oct. 30-31, pp. 344-354.

Kanbara, T., Miura, J., and Shirai, Y., 1993, "Selection of Efficient Landmarks for an Autonomous Vehicle." *Proceedings of the 1993 IEEE/RSJ International Conference on Intelligent Robotics and Systems*, Yokohama, Japan, July 26-30, pp. 1332-1338.

Kortenkamp, D., 1993, "Cognitive Maps for Mobile Robots: A Representation for Mapping and Navigation." *Ph.D. Thesis*, The University of Michigan.

Krotkov, E., 1991a, "Laser Rangefinder Calibration for a Walking Robot." *Proceedings of IEEE International Conference on Robotics and Automation*, Sacramento, CA, Apr. 9-11, pp. 2568-2573.

Kuipers, B. and Byun, Y., 1988, "A Robust Qualitative Method for Robot Spatial Learning." *The Seventh National Conference on Artificial Intelligence*, pp. 774-779.

Kurazume, R. and Nagata, S., 1994, "Cooperative Positioning with Multiple Robots." *Proceedings of IEEE International Conference on Robotics and Automation*, San Diego, CA, May 8-13, pp. 1250-1257.

Lebegue, X. and Aggarwal, J., 1994, "Generation of Architectural CAD Models Using a Mobile Robot." *Proceedings of IEEE International Conference on Robotics and Automation*, San Diego, CA, May 8-13, pp. 711-717.

Levitt, T., Lawton, D., Chelberg, D., and Nelson, P., 1987, "Qualitative Navigation." *Proc. DARPA Image Understanding Workshop*, pp. 447-465.

Lu, Z., Tu, D., Li, P., Hong, Z., and Wu, B., 1992, "Range Imaging Sensor for Auto - Vehicle Guidance Applying an Optical Radar." *Proceedings of the 1992 SPIE Conference on Mobile Robots,* Boston, MA, Nov. 18-20, pp. 456-465.

MacKenzie, P. and Dudek, G., 1994, "Precise Positioning Using Model-Based Maps." *Proceedings of IEEE International Conference on Robotics and Automation*, San Diego, CA, May 8-13, pp. 1615-1621.

Malik, R. and Polkowski, E., 1990, "Robot Self-Location Based on Corner Detection." *Proceedings of the 1990 SPIE Conference on Mobile Robots,* Boston, MA, Nov. 8-9, pp. 306-316.

Malik, R. and Prasad, S., 1992, "Robot Mapping with Proximity Detectors." *Proceedings of the 1992 SPIE Conference on Mobile Robots,* Boston, MA, Nov. 18-20, pp. 614-618.

McGillem, C. and Rappaport, T., 1989, "A Beacon Navigation Method for Autonomous Vehicles." *IEEE Transactions on Vehicular Technology*, Vol. 38, No. 3, pp. 132-139.

McKendall, R., 1988, "Robust Fusion of Location Information." *Proceedings of IEEE International Conference on Robotics and Automation*, Philadelphia, PA, April 24-29, pp. 1239-1243.

McVey, E., Drake, K., and Inigo, R., 1986, "Range Measurements by a Mobile Robot Using a Navigation Line." *IEEE Transactions on Pattern Analysis and Machine Intelligence*, Vol. PAMI-8, No. 1, pp. 105-109.

Ohya, A., Nagashima, Y., and Yuta, S., 1994, "Exploring Unknown Environment and Map Construction Using Ultrasonic Sensing of Normal Direction of Walls." *Proceedings of IEEE International Conference on Robotics and Automation*, San Diego, CA, May 8-13, pp. 485-492.

Parker, K., 1993, "'Bots Struggle to Learn Basics." *Manufacturing Systems*, Oct. 12, pp. 13-14.

Partaatmadja, O., Benhabib, A., Sun, A., and Goldenberg, A., 1992, "An Electrooptical Orientation Sensor for Robotics." *IEEE Transactions on Robotics and Automation*, Vol. 8, No. 1, pp. 111-119.

Pears, N. and Probert, P., 1993, "An Optical Range Sensor for Mobile Robot Guidance." *Proceedings of IEEE International Conference on Robotics and Automation*, Altanta, GA, May 10-15, pp. 659-664.

Roth-Tabak, Y. and Jain, R., 1989, "Building an Environment Model Using Depth Information." *Computer*, June, pp. 85-90.

Roth-Tabak, Y. and Weymouth, T., 1990, "Environment Model for Mobile Robot Indoor Navigation." *Proceedings of the 1990 SPIE Conference on Mobile Robots,* Boston, MA, Nov. 8-9, pp. 453-463.

Safaee-Rad, R., Tchoukanov, I., Smith, K., and Benhabib, B., 1992, "Three-dimensional Location Estimation of Circular Features for Machine Vision." *IEEE Transactions on Robotics and Automation*, Vol. 8, No. 5, pp. 624-640.

Santos, V., Goncalves, J., and Vaz, F., 1994, "Perception Maps for the Local Navigation of a Mobile Robot: a Neural Network Approach." *Proceedings of IEEE International Conference on Robotics and Automation*, San Diego, CA, May 8-13, pp. 2193-2198.

Schwind, G., 1994, "Controls Offer Non-Wire Guidance Without a Costly Premium." *Material Handling Engineering*, March, p. 31.

Shertukde, H. and Bar-Shalom, Y., 1988, "Target Parameter Estimation in the Near Field with Two Sensors." *IEEE Transactions on Acoustics, Speech, and Signal Processing*, Vol. 36, No. 8, pp. 1357-1360.

Singh, K. and Fujimura, K., 1993, "Map Making by Cooperating Mobile Robots." *Proceedings of IEEE International Conference on Robotics and Automation*, Atlanta, GA, May 10-15, pp. 254-259.

Xu, H. and Chi, X., 1993, "Calibration and Parameter Identification for Laser Scanning Sensor." *Proceedings of IEEE International Conference on Robotics and Automation*, Atlanta, GA, May 10-15, pp. 665-670.

Yoo, J. and Sethi, 1991, "Mobile Robot Localization with Multiple Stationary Cameras." *Proceedings of the 1991 SPIE Conference on Mobile Robots,* Boston, MA, Nov. 14-15, pp. 155-170.

Zheng, J., Barth, M., and Tsuji, S., 1991, "Autonomous Landmark Selection for Route Recognition by a Mobile Robot." *Proceedings of IEEE International Conference on Robotics and Automation*, Sacramento, CA, April 9-11, pp. 2004-2009.

SUBJECT INDEX

AUTHOR INDEX

COMPANY INDEX